BREWING WITH HEMP

THE ESSENTIAL GUIDE

BY ROSS KOENIGS

Brewers Publications®
A Division of the Brewers Association
PO Box 1679, Boulder, Colorado 80306-1679
BrewersAssociation.org
BrewersPublications.com

Proudly Printed in the United States of America.
10 9 8 7 6 5 4 3 2 1
ISBN-13: 978-1-938469-77-0
ISBN-10: 1-938469-77-1
EISBN: 978-1-938469-78-7

Library of Congress Control Number: 2022937388

Publisher: Kristi Switzer
Technical Editor: Luke Chadwick
Copyediting: Iain Cox
Indexing: Doug Easton
Art Direction, Cover and Interior Design: Jason Smith
Production: Justin Petersen
Chapter Page Illustrations: Cory Campbell

To the many people in my life who've inspired me to push toward what's next. You know who you are.

TABLE OF CONTENTS

NOTE FROM THE PUBLISHER

This is the second of two books exploring the intersection of cannabis and beer. Cannabis is the name of the plant grown primarily for two products: the stem fiber product hemp, and the intoxicating resinous product marijuana. Consequently, depending on the reason it is being cultivated, the terms "hemp" and "marijuana" are often used to denote the cannabis plant itself. However, it is important to maintain a distinction between the two. Marijuana is cannabis that contains more than 0.3% tetrahydrocannabinol (THC) by dry weight. Hemp is cannabis that contains not more than 0.3% THC by dry weight. THC is the compound responsible for the inebriating effect of marijuana. Marijuana remains illegal under federal law, though currently 46 US states, Puerto Rico, Guam, and the District of Columbia have made at least some allowances for medical and/or recreational use under their laws. Hemp, on the other hand, is legal under federal law and its use in brewing will be the subject of this book.

Despite "cannabis" encompassing both hemp and marijuana, it is important to note that over half of all drug arrests in the United States between 2001 and 2010 have been attributed to possession of marijuana. The majority of these arrests have disproportionately targeted the Black community, despite usage rates for marijuana among Black Americans being on a par with white Americans.[1] It's also important to acknowledge that there is also a history of

---

1 "The War on Marijuana in Black and White," American Civil Liberties Union, June 2013, https://www.aclu.org/files/assets/061413-mj-report-rfs-rel4.pdf.

vilifying the plant by playing upon anti-immigrant sentiment. This context led us to carefully consider the language we would use in working on this subject. Considering the history of word choice, we also needed to balance using language that offers the clearest understanding of cannabis in all its iterations.

Going forward in this book, we will be using the term *Cannabis* (capitalized, italics) when referring to the genus or species of that genus, and cannabis when referring to the plant in general or the plant's cultivation in general. We will use the term marijuana when referring to the processed plant product that has psychoactive properties, especially in the legal context. You may see the term "Marihuana" noted in specific laws or quotations and this reflects the spelling used in legislation. Lastly, we will use hemp when referring to the cultivated cannabis plant that has not more than 0.3% THC by dry weight.

The Brewers Association does not encourage illegal activity by anyone, including members of the brewing community. The reader accordingly should note that the possession and use of marijuana or its components (e.g., THC extract) remains a serious federal crime as of the date of this book's publication. Even for certain hemp by-products (most notably cannabidiol, or CBD), their use commercially in food and drinks remains subject to adverse federal action at this time, as the federal Food and Drug Administration has not yet recognized CBD and similar by-products as either dietary supplements or ingredients "generally recognized as safe." Nevertheless, many observers believe that the federal government will legalize marijuana and recognize food and/or supplement uses for by-products like CBD within the next few years. Moreover, interest in the potential for beverages infused with cannabis plant derivatives runs high in the alcohol beverage industry, and has attracted substantial investment from established players within the industry. As such, the Brewers Association perceives a very real impending need to educate the brewing community at large about this emerging and rapidly-changing subject.

This book will serve as a snapshot in time and every effort has been made to ensure it is up to date at the time it goes to press. But, as already highlighted, both the applicable laws and our knowledge of the various benefits and risks of cannabis use are changing rapidly. Readers should be sure to do their research before embarking on this journey.

Kristi Switzer
Publisher, Brewers Publications
March 16, 2022

ACKNOWLEDGMENTS

No one accomplishes big things alone. I would like to acknowledge those in my life that helped me on this journey. First, I would like to thank my wife, Amy, for her love and support through this process. I am beyond lucky to have you in my life. Next, I would like the thank my family and friends who encouraged me and lent me an ear to discuss ideas and to vent my struggles. You all inspire me.

I would also like to thank my friends and colleagues at New Belgium, as well as the friends I've made and colleagues I've met during my years working adjacent to the cannabis business. You all gave generously with your time and knowledge and I am forever grateful for your insights.

Finally, I would like to thank my publisher, Kristi Switzer, and the team at Brewers Publications for putting this book together. I am humbled by this experience and am grateful that you all believed in me. Thank you.

INTRODUCTION

My journey into cannabis happened somewhat by chance. Hemp researchers connected to Colorado State University visited the New Belgium taproom on an unremarkable afternoon in April 2015 and asked to speak to my boss, Peter Bouckaert. The US Congress had just legalized hemp cultivation for research purposes under the 2014 Agricultural Act and many in Colorado jumped at the opportunity to study this newly legalized plant. The researchers wanted to show Peter some of their plants, saying that they had many distinct aromas that might be of interest to us as brewers. Peter, famous for his curiosity and penchant for novel brewing ingredients, got excited. He grabbed a few of us, piled us into his car, and drove to the south end of Fort Collins to visit the greenhouses of a company now known as New West Genetics.

To be perfectly honest, when I first smelled the different plants I didn't really get it. I didn't really understand much about hemp at the time but I did know one critical thing: it's related to marijuana but doesn't get you high. The plants looked similar to marijuana and most of them had nice aromas, though some smelled quite awful. It wasn't until I dug a bit deeper that the light really clicked on for me. I had too many preconceived thoughts about cannabis. I figured it would be a cold day in hell before the government allowed us to put THC in beer, so exploring options in hemp would be a wasted effort—there was nothing interesting in cannabis if it didn't have THC.

I fell prey to what a lot of people think about cannabis: it's a plant that can get you high and nothing more. I'm pretty sure I can remember saying something to the effect of "Why would I put fake weed in my beer?" It seemed boring at the time and, to be fair, I think I was suffering from the bad information generally available. This poor-quality information comes from both proponents and opponents of cannabis; it reflects what happens when you demonize and criminalize something for so long. Most of what you end up with is colloquialisms that have little semblance to the truth.

From a legal perspective, the only difference between the two plants is the content of delta-9 tetrahydrocannabinol, more commonly known as THC. If a cannabis plant is at or below 0.3 percent THC by weight, it is hemp. If it's above 0.3 percent THC by weight, it is marijuana. Both are cannabis. Both plants are varieties of *Cannabis sativa* L. under the Linnaean classification system; therefore, despite common misconceptions, both hemp and marijuana are regarded by botanists as the same species. Over the course of a couple of generations in the wild, marijuana plants can become hemp and hemp plants can become marijuana.

What I came to find was that hemp has lots of interesting qualities to it and should be of great interest to brewers. When describing what hops are to their tour guests, most brewers cite as a piece of trivia that hops and cannabis are botanical cousins; I hope this book can give them some additional anecdotes to spice up their tour banter. The vast amount of research that has gone into understanding hops in the last one hundred years gives us the perfect lens through which to study cannabis. Brewers are uniquely suited to understanding and promoting this newly legalized plant. Many of the processing technologies we use for hop products are applicable to hemp, we have expertise in analyzing related biochemical constituents commonly found in hops and hemp, we have a shared expertise turning agricultural products into useful consumer products, and we understand and respect the power of intoxicants and can communicate the responsibility required to consume intoxicants in a safe and legal manner.

Since that chance encounter where I was introduced to the world of hemp, I've developed quite a bit of knowledge and expertise around how to brew with it. Since I focus most my time at New Belgium researching and developing hoppy beers, my role puts me in a unique position to translate that knowledge and skill base into a useful resource.

Cannabis is a tricky subject. As you will find out, some aspects of cannabis will feel very familiar to brewers, while others will feel foreign. It took a group of us about three years from that first moment in the greenhouse to introduce America's first nationally available hemp-flavored beer, the Hemperor. During that time, we were forced to navigate various legal pitfalls and a regulatory environment that didn't know what to do with such a product. At the end of it all, we felt like we were barely scratching the surface of the possibilities that marrying cannabis and beer held, so we proceeded to do further research and learn more from people in the cannabis industry. This book is a continuation of that work and, it is hoped, will make the subject a bit more accessible.

We were fortunate to be in Colorado, the state that most aggressively took advantage of the 2014 Agricultural Act to get hemp operations off the ground. We learned a lot in those early years from farmers and researchers and continue to maintain a close network of advisors to keep us informed of this rapidly evolving landscape.

This book, *Brewing with Hemp: The Essential Guide*, indicates that we will spend most of our time talking about hemp and only periodically talking about marijuana. The first reason for that should be obvious: this is a book for a brewing and beverage industry audience; you can add some hemp ingredients to your beer but you cannot legally add marijuana ingredients. The government may permit marijuana and beer one day, but considering we prohibit adding caffeine to alcohol it seems unlikely the government will allow brewers to add a powerful psychoactive to their products.

This book strives to provide high-quality information about cannabis for both professional brewers and homebrewers and explores the many ways hemp provides greater opportunities to brewers than marijuana. With that being said, the technical information presented in this book applies equally to both hemp and marijuana.

While researching this book, I was struck by the gulf between different sources of information about cannabis. Many research articles were written for a law enforcement audience that was more interested in stopping the spread of cannabis instead of seeking to understand it. Other research contained information that only MDs and PhDs would have a chance at understanding. On the non-academic side, many cannabis articles are so jargon driven that there is little information to be gleaned. With increasing legalization of both forms of cannabis, research is just starting to catch up and provide insights into the true nature of how our bodies respond to cannabis. As long as science rather than short-term business interests drives development in this field, we will continue to uncover definitively useful and therapeutic uses of cannabis, as well as understand how to properly educate consumers about its benefits and risks.

From that vantage point, I would like you to view this book as a snapshot in history. There are many people out there currently dedicating their careers to better understanding cannabis; considering where we are starting from, academically speaking, there is a lot that we still do not understand. Research on the subject is rapidly evolving, so some of the information presented in this book on this topic may become outdated. When compiling this book, however, we worked hard to verify information and avoid propagating any of the abundant misinformation about cannabis.

WHAT YOU SHOULD EXPECT OUT OF THIS BOOK

First and foremost, you should have a clear and factual understanding of the cannabis plant and its many useful components. Much of the content will focus on how to exploit the glandular trichomes of hemp, the hair-like structures within which the plant manufactures aroma compounds and cannabinoids. I will also present additional uses of hemp that should be of value to brewers, beverage manufacturers, and people generally interested in cannabis.

You will also learn how to evaluate and use cannabis in beer in a safe and legal way. I will review quality assurance systems that any brewer should use when qualifying a new ingredient for brewing. Additionally, this book should give you an appreciation for those quality systems in place in the hemp and marijuana industries that are generally very different from the quality systems in place for brewing ingredients. Cannabis plants can do some crazy things: for instance, they can pull heavy metals like lead and cadmium out of the soil, so you certainly would not want such plants anywhere near a glass of beer. In setting out a common set of quality parameters, I hope this book will be as much of an education for brewers as it is for growers of cannabis.

You should also gain an appreciation for how much we still do not know about this plant to help you evaluate various claims made by different people in the business. There are a lot of modern-day snake oil salesmen claiming everything from cancer cures to a newfound environmental utopia. Taking the most positive view of these claims, it is true that cannabis writ large shows great promise for many areas of human society. But many claims come from people with a profit motive and it can be challenging to ascertain what sources of information should be trusted.

This book will give you an understanding of US laws surrounding the cannabis business and how to navigate them successfully and legally. I sincerely apologize to my international readers for not providing more information about cannabis law globally—please consult your local regulations and laws before exploring this subject in your country. I will also do my best to outline what is still unclear in the law, especially as it pertains to brewers, and will provide suggestions on how we as a community can advocate for better laws while ensuring safety and consumer health. The snake oil claims also expose businesses to legal liability, so we will discuss how to avoid legal risk in labeling and advertising.

Finally, this book will help you appreciate the history of cannabis and have some fun learning about how humans have used this amazing plant for several millennia. The history of cannabis is intertwined with the development of human civilization, yet few appreciate just how omnipresent it has been throughout our history.

One thing you will not get out of this book is a robust discussion of the medical applications of cannabis, nor the interactions of cannabinoids on the human body. I am a brewer, not a medical researcher. This book is about how to use cannabis in beer and beverages for legal recreational purposes, not to discuss its pharmacological effects in humans. There are many more people far more qualified than me on the subject and I would much rather leave the discussion of cannabis as medicine to them.

Before we begin, it is important to ask a critical question: why did hemp not stay a viable commercial crop? It's easy to blame cannabis prohibitionists, but I think that misses a broader point. Hemp has wonderful properties suitable for all sorts of products, but it is a natural product. Natural products break down. The main reason why hemp lost its place in the world was because of the synthetic revolution. Nylon ship ropes will last for decades and have a higher tensile strength than any natural fiber. Synthetic materials can be designed to meet exceptionally strenuous specifications but have one significant drawback: they are terrible for the environment. Nylon's ability to resist decomposition also means that it is a significant environmental pollutant. Many synthetic materials come from oil or other hydrocarbons that pollute our land and water as well as contribute significantly to climate change when burned. Humankind's ability to create synthetic products is altering our environment so dramatically, we must now consider whether the true cost of these products is worth the outcome.

We are at the precipice of a global environmental catastrophe. According to the latest report from the International Panel on Climate Change (IPCC), human activities like burning fossil fuels drive increases in global temperatures, making the last four decades successively warmer than any decade that preceded it. If continued unabated, this warming will result in rising ocean levels, mass extinctions of plant and animal species, dramatically altered weather, crop failures, and more. In short, it will end human society as we know it today (Masson-Delmotte et al. 2021, 4–31). Should we fail to reduce our climate-changing emissions and pollution, humanity's future will be dramatically altered. Should we succeed, we will radically reorient society to achieve greater harmony with our natural world.

I would like you to think about cannabis in two ways. The first is to think of cannabis as a crop that can produce useful compounds and materials. The second is, if we subscribe to the idea of cannabis as a useful crop, how do we prioritize crops that have the greatest potential benefit to society? We need a better framework to assess the true value of our crops. You can use every part of the cannabis plant for literally thousands of applications—we should be seeking out crops that maximize utility and minimize waste.

SO WHY A BOOK ON CANNABIS AND BEER?

On the surface, it may not seem like a book on cannabis and beer makes a lot of sense. While occasionally the two are imbibed together, there's very little history of the two sharing the same glass. There are technical challenges that we will discuss to help illustrate why this has historically been the case. Again, this book will not talk about combining psychoactive marijuana with beer, because that is illegal. One should always use the upmost caution when combining powerful intoxicants that can produce unpredictable results. That does not mean, however, that studying the theory of infusing beverages with cannabis is an unrewarding pursuit. Adult beverages—and beer specifically—offer such diversity these days that alcohol is no longer a requirement to be welcomed into the beer tent. The rise of the craft brewing movement has reinvented how we see beer and that reinvention has spilled over into numerous other beverages.

When you look deeper, cannabis and *craft* beer makes intuitive sense. Both industries arose from federally outlawed activities and were pioneered by a passionate group of home tinkerers. It's easy to forget that homebrewing was first legalized in 1978 and was still illegal in certain states as late as 2013. Both industries rely on farming and agriculture and have a vested interest in seeing that farms are efficient and economic. Both industries share a history of federal prohibition, albeit they did not share that history simultaneously. Both industries have complex regulations and require high levels of knowledge to successfully navigate the legal requirements to operate a commercial business selling either alcohol or cannabis. Both industries focus on hedonism and pleasure, albeit cannabis use is split between both recreational and medicinal. On that last point, it's important to remember that those two aspects of the commercial cannabis business never mix; when discussing alcohol combined with any cannabis derivative, remember that it will be regulated as a recreational substance that legal adults choose to imbibe.

Obviously, cannabis is still partially outlawed in the US and throughout the world. Attitudes, however, are changing quickly. As increasing numbers of states and countries legalize both hemp and marijuana cultivation and see the substantial societal and economic benefits resulting from cannabis legalization, more governments are rethinking their positions. Likewise, public attitudes toward cannabis legalization have dramatically changed in the last 15 years. According to one Pew Research study, support for cannabis legalization more than doubled, from 32 percent of the population in 2005 to 67 percent in 2019 (Daniller 2019). Much of this change has been driven by the attitudes of younger citizens, who are increasingly holding the levers of power. There seems to be an inevitability about the increasing legality and availability of cannabis throughout much of the world.

We've seen that cannabis legalization in individual US states can produce positive results while minimizing negative results, such as decreased violent crime associated with drug trafficking while not raising marijuana use rates amongst underage children (Gabrilova, Kamada, and Zoutman 2019, 375; Anderson, Hansen, and Rees 2019, 879), but lack of federal oversight undermines the potential of the business. As you will see, there is a lot of variability in how cannabis law is applied. In fact, a lot of enforcement of cannabis law contradicts other agencies. Absent federal legalization and the development of a robust national regulatory framework, navigating the cannabis space presents countless obstacles.

As legal barriers are dismantled, it is imperative that high-quality information about cannabis is disseminated. While there is a wealth of excellent information coming from the academic community, it is not very accessible to the average reader. As a result, there is a lot of cannabis writing that contains half-truths, anecdotal evidence, and insider jargon. Some of this derives from the fact that the capitalist system disincentivizes the sharing of useful information. Cannabis companies' unwillingness to share information is logical and, sadly, good business. This approach is necessary to delay the appearance of knock-off products and is a prerequisite to patenting any technology. A robust exchange of knowledge and information would serve producers and consumers better in the long term and make cannabis more accessible to new consumers. By and large, craft brewers know this struggle and took a different approach: our industry has, for the most part, deliberately chosen to openly share intellectual property rather than horde it.

I genuinely believe that the craft beer community has a lot to offer the emerging cannabis-infused beverage business and this is a driving force behind this book. I enjoy asking other craft brewers, "What does craft beer mean to you and why is it important to you?" I'll do my best to give my own answer, although admittedly it changes through the years. The ethos of craft beer is equal parts a mindset and a group of products. Craft strives to make things better: better tasting, better ingredients, better for you. Craft seeks to create products the right way, which usually means the hard way. It honors the ingredients, it honors the maker, and, by doing so, it honors the consumer.

There is a profound lack of information available on developing cannabis beverages. This book is a first attempt to address this problem and to highlight areas where more research is needed. My hope is that you think about the ideas I present in the book with the same voracity as you would about how to make your next beer. When you think about how a beer gets made, there are a lot of components from different parts of the world that must come together. We as brewers rely on a robust supply chain that has decades, if not centuries, of experience in food quality and regulatory compliance. When it comes to the cannabis space, there is a wide variation in experience and expertise in these realms. It is imperative that anyone looking to work with cannabis applies an independent, critical mindset to ensure the quality of their product, because deficits in quality and regulatory compliance can have terrible consequences. We saw this in the fall of 2019 when manufacturers of marijuana vaporizer cartridges added vitamin E acetate to their products, resulting in the deaths of 54 people and sickening of over 2,500. It is true that many of those deaths were attributable to black or gray market sellers, but there were several licensed manufacturers who sold products using the same deadly formulations. If cannabis is to be ever legalized federally, this type of criminal negligence needs to be eradicated. Paradoxically, we cannot eradicate this criminal negligence without robust regulatory standards.

Cannabis has the potential to bring a lot of good to the world. I think the craft beer community has a large role to play in shaping the moral arc of the burgeoning cannabis industry. Most craft business models focus on creating value up and down the supply chain. Since a higher price is paid for the end product, it allows upstream suppliers to get a more sustainable price for their raw materials. This creates a much more sustainable business structure as opposed to more traditionally commoditized business models. If you need an example of this, look no further than the Yakima Valley of Washington State.

Before about 2000, plus or minus a few years, hop growers in the valley were shedding acreage and consolidating operations at a rapid pace. Most growers were not earning enough money to reinvest in operations, let alone pay their existing bills. The large brewers and hops brokers had squeezed much of the margin in the highly commoditized hops market that focused on alpha acid levels and little else; there just simply wasn't much money to be made at any part of the value chain. Lo and behold, the craft beer phenomenon started; brewers came along looking for high-value aroma hops and the story of Yakima changed dramatically. Farmers were able to reinvest in their operations, start and fund research and breeding programs, figure out how to harvest more efficiently, and, most importantly, make money.

It is through the example of the hop growers in Yakima Valley that I think we should all consider how to build a sustainable cannabis business. Make no mistake, the business is not even close to being built. New technology and infrastructure will need to be put in place and markets will need to be created. If we target the wrong thing too early, we risk seeing a collapse of markets and support for innovative new products. Thus, we need to think about how we capture value from low-hanging fruit; what that is will look different to different people, which is a good thing. So, working from my position of strength, I am going to talk quite a bit about what I know: how to design beers using hemp. Most of what I will discuss will focus on flavor; however, I will also lay out the theory and technologies of how one would go about infusing cannabinoids into a beverage.

I hope you enjoy this book and you get something useful out of it. I am excited about what the future of cannabis beverages holds. I hope that this book can play a small part in shaping that future.

1

A BRIEF HISTORY OF CANNABIS

When starting with a history of cannabis, some definitions are essential. There is no fundamental difference genetically or morphologically between marijuana and hemp: they are both cannabis. The only difference between them is THC concentration.[1] Humans have imposed this distinction between marijuana and hemp over time as they domesticated the plant. Currently, the US federal government places that distinction at 0.3 percent THC by weight.[2] Does your cannabis plant have 0.30 percent THC by weight? You've got hemp. Does it have 0.31 percent? You've got marijuana. Keep in mind when reading that I use these three terms deliberately. When I say **cannabis** or ***Cannabis*** (the latter being the genus to which cannabis belongs), I am writing about something that applies to both marijuana and hemp. If I am using either **hemp**

---

1 THC stands for delta-9 tetrahydrocannabinol, the psychoactive compound that is commonly found in the cannabis plant.

2 Domestic Hemp Production Program, 7 C.F.R. § 990.1 (2021).

or **marijuana,** the information pertains directly to either the specific regulated industry or to characteristics that force the legal distinction.

While this is a brief overview of the history of cannabis, I recommend several excellent books for well-researched accounts of this fascinating subject. Robert Clarke and Mark Merlin's *Cannabis: Evolution and Ethnobotany* explores the subject from a multidisciplinary approach; Martin Booth's *Cannabis: A History* examines the history of human use of marijuana; Martin Lee's *Smoke Signals: A Social History of Marijuana* looks at marijuana and hemp's history and illegalization over the last 150 years; and, finally, Ernest Abel's classic *Marihuana: The First Twelve Thousand Years* is a must-read for its historical significance and foundational research.

THE EARLY EVOLUTION OF CANNABIS

I start this history of cannabis at the beginning of its evolution. While that may seem obvious, it's especially important considering the vast pressure ancient humans subsequently placed upon cannabis and their direct actions that spread the plant beyond where it had originally adapted to. The divergence point in history where cannabis ceased to be a wild plant and became domesticated is unclear. What is clear is that the history of cannabis is inextricably linked to the rise and development of human civilization.

The earliest form of flowering plants evolved sometime during the Lower Cretaceous period, about 125 million years ago. Flowering plants of the taxonomic family Cannabaceae—along with all modern flowering plant families—are thought to have formed in the early Upper Cretaceous about 100 million years ago. Archaeologists have not found a fossilized record of any proto-cannabis species to date, so much of what those early plants could have looked like is subject to conjecture. What is understood is that the two best-known genera of the Cannabaceae family—*Humulus* (hops) and *Cannabis*—diverged from an unknown common ancestor around 28 million years ago. Archaeologists and archaeobotanists studied fossilized pollen samples from around the globe and used sophisticated models to track and date them. A challenge in this type of study is that *Humulus* and *Cannabis* pollens are morphologically similar, thus, researchers had to look at other pollens within the samples as markers to help distinguish what was in each sample. Through the study of these pollens at multiple sites, along with analyzing their geographic location and extrapolated environments, they have provided compelling evidence for the evolutionary timeline of Cannabaceae and when *Humulus* and *Cannabis* diverged from their common ancestor.

McPartland, Hegman, and Long (2019) presented the best evidence to date of *Cannabis* evolving from its ancestor on the Tibetan Plateau around 28 million years ago. They additionally deduced that the genus's center of origin was around Qinghai Lake about 1,300 kilometers northwest of Chengdu, China. *Cannabis* then dispersed west, likely through seed dispersal by a combination of wind and rivers, reaching Europe around 6 million years ago. It also eventually traveled into eastern China around 1.2 million years ago. The study also noted the relatively late appearance of *Cannabis* pollen in India, which showed up only around 32,000 years ago (McPartland, Hegman, and Long 2019, 691)

McPartland and Guy (2004, 71) put forth a theory that humans and *Cannabis* coevolved together with the human development of the endocannabinoid system, which will be discussed further in chapter 5. However, many animals have similar endocannabinoid systems, so it is not a unique human adaptation. It is plausible that our early hominid ancestors in Africa (and later in Asia) ingested cannabis and then identified and selected the most potent varieties. It's equally plausible that in the course of history humans were so attracted to the nutritional and psychoactive properties of cannabis that they sought it out wherever they could until they were able to successfully grow it. Considering *Cannabis* was present growing in the wild in Europe and Asia when *Homo erectus* reigned, it is certainly plausible that our hominid ancestors, already capable of manipulating fire and using tools,[3] could have foraged wild *Cannabis* and possibly spread it to a broader geographical range during the period stretching from 1.5 million to 200,000 years ago. Despite little evidence in the archaeological record, it is almost certain that *Homo erectus* encountered wild *Cannabis* at some point in the nearly one million years during which both existed in the same relative geographical area.

In the fascinating book, *Cannabis: Evolution and Ethnobotany*, the authors imagine a time when modern *Homo sapiens* spread across most of the globe and came into contact with cannabis plants for the "first time" on the steppes of Central Asia. They describe a "camp follower" theory of initial domestication, where humans played a role in the planting and dispersal of *Cannabis* but did not immediately grow it in a way that would be recognizable as modern agriculture (Clarke and Merlin 2013, 1–6). Clarke and Merlin's theoretical band of nomadic hunter-gatherers could have existed at any time since the emergence of *Homo sapiens*, upward of 200,000 years ago. These hunter-gatherers would have

---

3 Richard E. Leakey and Alan C. Walker, "*Homo erectus* unearthed," *National Geographic*, November 1985, 626.

spent much of their time near rivers for access to fresh water and encountered wild cannabis there. By late summer to early fall, they would have found that the cannabis plants fruited and, once their natural curiosity and instinct to forage kicked in, they would have eaten them. Not only did the humans not fall sick or die from eating the plant, they found cannabis to be delicious and highly nutritious. These hunter-gatherers perhaps set up camp for a few weeks around the fruiting cannabis plants so they could feast on the bounty before them.

At some point the nomads would move on, taking excess seeds for the journey ahead and remembering the spot on the river so they could return the following year. Naturally, a few seeds would have hit the ground and sprouted the following spring. When the nomads returned to the site of the original discovery, to their delight they found the cannabis plants thriving, fertilized by the organic waste of the camp from the year before. This process would repeat every year, until the nomads discovered more stands of cannabis, possibly through the wayward seeds they dropped in years past.

Eventually, Clarke and Merlin posit, the nomads would have cut down the plants to more efficiently harvest the seeds. By dragging the plants back to camp, they further spread seeds for growth; once they harvested what seeds they could, they would throw the waste plants into a rubbish heap. Any remaining seeds would sprout and thrive in this nitrogen-rich environment, resulting in the nomads having self-sustaining stands of cannabis near their camps. When the nomads no longer needed to travel so far to gather their food, an incipient form of agriculture was born.

As this group's cannabis became more abundant, Clarke theorizes its members became increasingly curious as to what other uses could come from their crop. Since knowledge was only passed orally, new technologies developed slowly. But, eventually, the group figured out methods to press the seeds into oil, which could be used for cooking or fashioned into a crude soap. While these nomads mostly wore animal skins, they required plant-based fibers to hold the skins together, and soon discovered that they could peel the bark off the stems of their cannabis plants and make a suitable weaving fiber. Further experimentation would show that if they let the stalks sit in a pond, or in a spot in the river with a slack current, they could improve the quality, strength, and durability of these fibers.

Finally, at some point this group of nomads—perhaps now living part of the year in a dedicated settlement near their cannabis crop—would have decided to have a large community gathering one evening. Setting a large fire in the middle of the village, they would have sat around to eat, telling stories and socializing. As the fire died down during the evening, they might have found themselves short of firewood, which would lead them to their rubbish pile and a decision to burn some of the dried plant matter from their cannabis crop. With smoke rising from the fire, the villagers would begin to feel strange. The first cannabis "party" had begun.

While hypothetical, Clarke and Merlin's plausible story illustrates the common themes around how humans came to understand the utility of *Cannabis*: its nutritious seeds, its durable fibers, and its powerful pharmacology. Since then, human civilization has oriented itself around harnessing these three products, which informs how *Cannabis* affected historical events for millennia.

Cannabis's history is similar to its role in civilization today. Its importance has ebbed and flowed through different cultures, with many waves of discovery, exclusion, and rediscovery. Great joy and sorrow exist in its history, often simultaneously. Many societies have sought to control the use of cannabis as a recreational drug, recognizing the threat that a mind-expanding substance can be in a population that the powerful seek to control. This notoriety has been its greatest asset and its greatest downfall—countercultural elements have harnessed the notoriety to effect cultural change, but by being a symbol of counterculture movements cannabis has often struggled to gain mainstream acceptance.

ANCIENT CULTIVATION AND USE OF FIBER

Myriad pieces of evidence show that modern humans (*Homo sapiens*)—who evolved roughly 200,000 years ago—played a singular role in the vast spread and eventual domestication of cannabis. Part of the challenge to getting an accurate historical record of early cannabis use is that most of the material naturally decomposes over time. Archaeologists must therefore look for sites and artifacts that enable them to estimate *when* what they are observing happened. This often involves looking at pottery shards that may have had cordage imprints or, if they are lucky, they will have a site with climactic conditions that deter decomposition, such as a desert or a bog. In these fortuitous instances, archaeologists can directly prove the presence of cannabis at the site and theorize its importance to past civilizations. Most of what follows in this section is gleaned from the *Great Book of Hemp* (Robinson 1996).

The earliest direct evidence comes from a Neolithic site in present-day Taiwan, dating back almost 12,000 years ago. Pottery shards recovered from the site show it was decorated with cords, and indirect evidence of the raw materials used during that period make it likely that the cord was made from hemp. Along with the pottery, archaeologists discovered a stone beater that additional evidence indicated was used for the processing of hemp fiber (Robinson 1996, 103). It is definitively proven that humans were planting cannabis throughout the rise of the domestication of plants and animals, roughly around 10,000 years ago. Some historians and anthropologists theorize that the utilization of plant fibers to create tools and clothing had a significant influence on societies transitioning from hunter-gathering to farming, domesticating livestock, and living and working in permanent cities.

In addition to artifacts and biofacts, the advent of written language systems means there is also a historical archaeological record to help track the presence of cannabis throughout ancient civilizations. The first discovered written language emerged roughly 6,000 years ago in the city-state of Uruk, in what is now Iraq. The use of writing systems then exploded throughout the world's early civilizations, becoming commonplace within 2,000 years of their invention. As writing systems proliferated around the globe, cannabis's story becomes clearer and more dynamic. Direct written evidence of cannabis use came considerably later, first being mentioned in Egyptian texts roughly 3,500 years ago. Shortly thereafter, we find the first documented references to cannabis in several societies.

Early humans more than likely used hemp to stitch together animal hides or other crude forms of clothing, which led to innovations for efficient fiber extraction, some of which are still in use today. The most direct way is to peel the outer fibers off the base of the stalk, either freshly after cutting the plant down or after it dries. Fibers processed this way, however, are gummy and hard to use, so humans quickly realized that fiber quality could be vastly improved through a process called *retting*. Meaning "to rot," retting is a method that enables the bacterial and enzymatic break down of certain plant fibers. The traditional method involves soaking hemp fibers in water and allowing natural bacteria and fungi to break down the rigid stalk, freeing long fibers for further processing. In certain climates, farmers can "dew ret" hemp stalks by piling them and waiting for rain to sufficiently saturate them.

Throughout most of history—and currently in certain parts of the world with lax environmental regulations—farmers would submerge their hemp crop in a nearby ditch, creek, or pond for retting. "Water retting" in this fashion leads to high levels of water pollution and the bacterial fermentation causes such low oxygen levels in the water that it kills wildlife living there. China, unfortunately, still practices this type of retting for convenience and cost, and as such controls most of the hemp fiber and textile trade globally (Zhang et al. 2008).

The development of natural fiber technologies played a large role in the advancement of organized human civilization. Strong, durable fibers allowed humans to spread to more diverse environments because they could better manipulate and control the natural world. Cordage and spun natural fibers allowed humans to fasten, trap, hold, and carry things. It is no accident that the advent of fiber-based cordage technologies coincided with animal domestication, further allowing early humans to shape their environment to their advantage. Over time, the complexity of these cordage technologies grew more sophisticated, allowing for the specialization of work and eventually aiding in the creation of complex, cooperative societies.

No single plant species provided all the natural fiber for every ancient civilization. Initially, early humans sought fiber technologies to fasten animal furs, and they likely experimented with a variety of fibers before landing on a few that grew abundantly in their environment and worked well for this purpose. Early humans adopted hemp where they found it, most often in Asia, Russia, and Northern Europe. Flax became the preferred fiber of the Mediterranean, sheep's wool in Mesopotamia, and cotton in India and South America. As civilizations became more complex and trade emerged, the preferred fibers of any locality eventually gave way to the use and proliferation of multiple different fibers suited to specialized needs.

In ancient Egypt, the use of flax was far preferred over other natural fibers, especially for creating linen. Cultural preferences prioritized flax fibers for clothing and ornamental applications, but as the time went on Egyptians incorporated hemp cordage into heavy labor sometime between 5,000 and 3,000 years ago. Archaeologists have found evidence for *Cannabis* cultivation in the Mediterranean from as early at the seventh century BCE; the original Hebrew Bible and the Talmud mention hemp rope in the construction of great temples. It is also likely that hemp cordage played a major role in the construction of the Pyramids; unfortunately, there is little archaeological evidence of this to date, but there are plenty of examples of the use of hemp rope in Egypt at that time to support the theory. Additional evidence shows that the ancient Egyptians also knew of the medicinal properties of hemp; texts from as long as 3,500 years ago mention the use of hemp to alleviate pain and inflammation from a host of diseases (Russo 2002, 6). Egypt is by no means unique, in fact, most ancient civilizations adopted hemp fiber to some degree, especially when it came to industrial, military, and sea-faring applications.

ORIGINS OF DOMESTIC FIBER USAGE IN CHINA AND EAST ASIA

The best documentation of early hemp fiber usage is in China and East Asia, where arguably great advances in hemp fiber technology occurred far earlier than in other civilizations. There are countless examples of hemp fiber being used for cordage, paper, textiles, and clothes throughout Asia, and it was one of the most dominant crops for multiple millennia. In fact, the early Chinese regarded hemp so highly that they referred to their nation as "the land of mulberry and hemp" (Abel 1980, 5). Mulberry, whose leaves feed silkworms in the production of silk fibers and cloth, often represented the noble and aristocratic members of Chinese society, whereas hemp often represented the peasant and working classes (Li 1973, 441).

Silk fibers compare quite well to hemp fibers in their durability, strength, and elasticity but its production could never be scaled in a way to allow its adoption as a universal fiber in ancient China. Thus, hemp, along with other cordage fibers, stepped in to fill the applications where silk was too expensive to use: rope and cordage, utilitarian clothing, stitching yarn for clothes and armor, sandals and shoes, banners, paintings, and embroidery (Abel 1980, 4–6; Booth 2003, 20–22). Spinning hemp fiber was a seasonal cottage industry for rural Chinese people, often relegated to women of the peasant classes; this remained the case for thousands of years (Li 1973, 441).

Hemp garments also played important roles in mourning, a tradition that continues today in parts of China, Korea, and Japan. Wearing custom hemp clothes was a form of conferring respect for the dead; the color, type, and specifications for the cloth date back to funerary traditions of Confucianism. Mourners could wear nothing but their hemp garments for up to three years, depending on their relationship to the deceased (Clarke and Merlin 2013, 271).

The significance of hemp clothing in death and funerary rights may have been passed on to several Pontic Steppe cultures, including the Scythians, a nomadic people that lived on the Pontic Steppes of Russia and Mongolia between 7,000 and 2,000 years ago. The Scythian Empire existed between the eighth and third centuries BCE; it's referenced frequently in cannabis history because of the Scythians' importance as traders and connectors on the Silk Road. The Greek historian Herodotus made mention of the Scythians' use of hemp textiles and cordage in their lives and funeral rituals, as well as their zeal for inhaling the fumes of cannabis seeds (Godley 1921–25, 2:272–275). The Scythians' reputation as cannabis lovers can be traced through much of the history of ancient cannabis, and several earlier Pontic Steppe cultures may have played a large role in spreading the use of marijuana to Mesopotamia and the Mediterranean.

DEVELOPMENT AND PROLIFERATION OF PAPER

The history of cannabis is intertwined with the invention of one of the most important technologies ever developed: paper. Before the advent of paper, most writing systems were engraved onto solid media such as rock, wood, clay, or shells. According to legend, an official of the Chinese royal court, Ts'ai Lun, invented paper to lessen the physical burden of delivering documents to the emperor (Abel 1980, 6). Because they were written on bamboo slips and wooden tablets, servants of the court would daily haul hundreds of pounds of documents to the emperor to conduct business. Ts'ai Lun explored using silk as a paper source but silkworms could not produce enough to meet the writing demands of a growing empire.

The original paper was likely a composite of various plant fibers. While hemp often gets credited as the sole source of paper fibers, a combination of bamboo, mulberry, rice straw, rattan, and other plant matter probably comprised the first papers (Clarke and Merlin 2013, 187). With increasing availability of paper products, Chinese papermakers found creative solutions for sourcing additional hemp-based feedstock for paper, including rope ends, rags, and worn out fishing nets (Pan 1983, 179)

Trade of hemp paper between China, Korea, and Japan predates trade through the Silk Road. Starting approximately 2,500 years ago, hemp paper and cordage products spread through the Silk Road trading routes to Tibet, India, the Middle East, North Africa, and Europe. Those routes spread not only the paper itself, but also the underlying technology to produce it locally.

USE OF HEMP FIBER FOR TECHNOLOGY AND WARFARE

There is more than a little irony in the fact that cannabis, a crop currently associated with hippies and peace-proliferating ideologies, has a long and storied history in waging war. Hemp fibers produce strong and resilient cordage products, which played a large role in many civilizations' ability to haul heavy equipment and transport goods and people. Hemp's passive role in warfare includes rigging and sailcloth for warships, saddles and hauling mechanisms for utilizing animals in warfare, and ropes that powered siege weapons, including catapults.

One of the most important military inventions that came from the discovery of hemp fiber's elasticity and durability was the bow and arrow. The Chinese were first credited with developing hemp fibers as a string material for their bows as they looked for a more durable alternative to bamboo fibers. Using hemp allowed them to invent more powerful technologies such as the longbow and the crossbow (Clarke and Merlin 2013, 146). In fact, producing sufficient quantities of bowstrings was such as priority that "Chinese monarchs of old set aside large portions of land exclusively for hemp," arguably making hemp the first-ever production war crop (Abel 1980, 5).

Another example are the catapults developed in China, where the powering ropes were often hemp cordage and leather wrapped together. The two materials allowed for operation in any weather condition and provided flexibility, strength, and elasticity in the launching mechanism. Additionally, hemp was used in the making of early incendiary devices launched by catapults. Large objects were wrapped in hemp fibers, dunked in resin, wax, and oil and then set ablaze before launching (Franke 1974, 169). Crueler means of inflicting destruction sometimes involved tying burning bundles of hemp to oxen or birds and releasing them into enemy towns to cause panic and chaos (Clarke and Merlin 2013, 147).

In Japan, military nobles—the samurai—also placed cannabis in very high esteem, prizing hemp fiber for their weapons and capturing its significance in art and cultural practices. Additionally, members of the Japanese military would traditionally wear hemp clothing into battle for both practical and symbolic purposes. Using hemp in military garb is by no means unique to that period, there are countless examples where hemp is privileged as a preferred military cloth. Any time you find a historical culture that utilized hemp, it served a useful role in warfare. Many great armies that relied on horsepower, from pulling chariots to transporting heavy equipment, required abundant hemp cordage for their war efforts.

With greater trade, communication, and exposure to different civilizations through the Silk Road, many hemp technologies spread throughout Eurasia and Africa. It is certainly possible that other civilizations, especially in Europe, developed hemp fiber technologies independent of knowledge sharing by the Chinese. There is a vast amount of evidence that hemp cultivation in Europe predates any exposure to technologies brought by the Silk Road (Mercuri, Accorsi, and Bandini Mazzanti 2002, 265). But whether the Chinese directly spread civilian and military hemp technologies is beside the point. The point is that at any given time in the historical and archaeological record involving a structured or complex culture, human societies stretching from Europe to Japan were cultivating hemp fibers to power technologies that built and conquered civilizations.

HEMP FIBER AND ITS ROLE IN GLOBAL EXPLORATION, FISHING, AND MARITIME APPLICATIONS

When ancient cultures realized that hemp fiber was particularly useful at sea, hemp truly became a global commodity. Ships allowed armies to move quicker and in greater numbers, and hemp turned out to be the ideal fiber for making durable sail cloth.

Humans understood very early on that hemp fiber had special properties that resisted decomposition in water. That understanding began with the development of retting technology and flourished as humans discovered the many maritime applications of hemp fibers. These applications led to multiple complex economic systems that arose from maritime technologies, including global exploration and trade.

Initially, most maritime applications revolved around fishing. Hemp fiber's relatively high resistance to water allows it to be submerged in both fresh and saltwater for extended periods of time and still be functional. Eventually it will degrade and break down, but hemp is superior to most natural fibers in terms of water resistance and durability. Thin, woven hemp fiber strands made durable fishing lines and expanded early humans' ability to feed themselves. The invention of woven fishing nets allowed humans to capture larger quantities of fish and feed larger groups of people. At some point, humans made crude rafts, threshed together for stability and floatation. The age of sailing was born.

The first known sailing culture arose in Southeast Asia sometime between 10,000 BCE and 6,000 BCE in and around what is now Taiwan and Indonesia (Gray, Drummond, and Greenhill 2009, 480). Members of the Austronesian language culture, which covered the region from present-day Malaysia to the Polynesian Islands, were the main inhabitants of this area. The Austronesians established an extensive maritime trading system that eventually included China, India, and Southeast Asian cultures. Ostensibly, early learnings from the Austronesians gave rise to Chinese sailing technologies and eventually spread to seafaring cultures around the Indian Ocean, including those in Africa and the Middle East (Manguin 2016, 54).

Moving outside Asia, Egyptians are often credited as the first Mediterranean sailing culture, with ships appearing in hieroglyphs dating back to 5,000 years ago. Whether ancient Egyptian sailing technology arose independently or in conjunction with other seafaring cultures is unknown; what is known is that hemp cordage and fabrics were essential to constructing the ships and the sails that powered them. Hemp cordage held together all types of ship designs and provided the essential rigging systems to hoist sails. In fact, according to Clarke and Merlin (2013, 161), the term canvas is derived from the word cannabis, claiming

> the etymological derivation of canvas being from the Anglo-French word canevaz, which comes from the Old North French *canevas*,

which in turn is assumed to have been derived from the Vulgar Latin *cannapaceus* ("made of hemp"); the Latin *Cannabis* is believed to have its origin in the Greek *kannabis* ("hemp").

Hemp sail cloth became a prized commodity throughout Mediterranean civilizations, with every major culture promoting hemp cultivation and production for ship building.

Romans spent centuries breeding hemp for high-quality, durable fibers. This reputation continues to this day in Italy, where hemp fiber from certain provinces is still prized. The Roman Empire was arguably built upon its mastery of commerce, combined with a strong military to keep markets functioning in an orderly way. The Roman hemp fiber industry kept ships sailing, which kept the empire functioning. Hemp also improved ships and allowed them to travel greater distances by being a crucial component in a waterproofing product called oakum. Created by taking specially arranged hemp cordage and impregnating it with tar, ship builders would pack oakum between ship joints, dramatically improving the watertightness of the ship. During construction, additional hemp cordage would compress these oakum-packed seams until they dried and sealed.

As the Roman Empire expanded, so did the importance of planting hemp throughout Europe, the Middle East, and Africa. Certainly, hemp cultivation in those regions predated the Romans, but the empire provided a formal marketplace for hempen goods. Most hemp cultivation previously existed in service of a town or even a single farm; the Romans enabled it to become a dominant industry. Large-scale production of hemp spread throughout Europe, where it became crucial for the civilizations that eventually supplanted the Roman Empire.

After the fall of imperial Rome, cultivating hemp remained a priority for many civilizations, especially those that traveled by sea, which included the Viking tribes of Scandinavia, who explored large tracts of the Atlantic as far as North America. Hemp fiber cordage played an integral role in the construction of sailing ships and allowed those civilizations to explore and conquer territories.

A frequent theme in the history of hemp and world exploration is its versatility for travelers. Madagascar, for example, had few global trading partners save for African nations. At some point cannabis was introduced, potentially because a seafaring people were marooned on the island. Clarke and Merlin (2013, 128) suggest that many sailors carried cannabis seeds in case they got stuck somewhere so they could possibly grow the raw materials needed to fix their ships and carry on their voyage. This dissemination theory comes in part from the well-documented practices of European kingdoms seeking to explore new parts of the world, which formally started in the fifteenth century. Spain, Italy, England, France, and the Netherlands all sought to expand their reach and capture untold treasures purported to be in the New World. Centuries of improvements in hemp technologies meant it was possible to build ships capable of sailing for months or even years at a time to explore uncharted lands in North and South America. Among the cargos carried by these explorers was plenty of hemp seed to plant in the newly discovered territories. New hemp plantations in what are now Brazil, Cuba, Mexico, and Haiti provided additional cordage material for local ship repairs and replacement supplies for long voyages. Extra cordage could be traded with the native peoples or sent back to European kingdoms.

This is also the beginning of a sad and tragic period in the history of cannabis, with ramifications that stretch on into today. The exploration period ushered in the age of colonialism: a period marked by the enslavement and subjugation of millions and the institutionalization of racism and oppression. Hemp fiber cordage unfortunately enabled colonial governments to enact these institutions and it became one of the primary crops that provided the economic incentives to make these systems operate. The plant that brought humanity the sail and the saddle also brought the whip and the noose.

Hemp became a staple crop in any area of North or South America that was under the control of European colonial powers. While perhaps not featured as prominently as crops like tobacco, cotton, sugar cane, or even tomatoes in history lessons, hemp (along with flax) had an omnipresent role in agricultural plantations of the New World colonies. The lack of discussion around the historical importance of hemp in the New World can be explained somewhat by its utilitarian role, but a strong argument could be made that its omission from much of history is due to what hemp fiber represented. Hemp cordage powered the ships that enabled the trafficking of human beings for subjugation. Those enslaved humans then were forced to grow hemp, which was turned into tools that were used to enforce their subjugation and to enslave others.

This idea can best be summed up by the history of hemp in Kentucky. Kentucky became famous as a hemp cultivation zone during the late 1700s and early 1800s, a reputation that continues today. The rise of the cultivation of hemp in Kentucky, however, also coincides with the proliferation of slavery throughout the state. The hemp fibers produced in Kentucky enabled early, free Americans to travel west and explore. Those fibers made the canvas that covered the travelers' wagons, tacked the saddles of their horses, and pulled them to a new future. Those fibers also took more ominous roles in the form of the hangman's noose, used to issue frontier justice and to terrorize communities of color. In fact, many in the West described the noose as the "hemp collar" (Lee 2012, 19).

At the height of hemp fiber's popularity in the mid-nineteenth century, two key technologies came along to disrupt it: the steam engine and the cotton gin. While both of these technologies were invented in the eighteenth century, their mass adoption did not occur until the latter half of the nineteenth century. The steam engine eventually took away altogether the need to propel ships by sails and the cotton gin enabled cotton farms to produce cheaper and better-quality cotton textiles than hemp. Within a few decades of their mass adoption, these technologies disrupted the global hemp supply chain and hemp never fully recovered as a major crop.

HISTORICAL USES OF HEMPSEED

Unlike cereal crops grown mainly for their starch-laden grains, hempseed is an oilseed crop whose main use is for the extraction of oil; like many other oil seeds, they can also be eaten whole, either raw or roasted. Hempseed contains roughly equal quantities of carbohydrate, protein, and fat, and the full spectrum of omega-3 and omega-6 fatty acids that are essential for human metabolic functions and, as we will discuss later, those of yeast. Thinking back to Clarke and Merlin's story of *Homo sapiens*' theoretical first contact with *Cannabis*, they theorized those early humans first interacted with cannabis plants by eating the nutritious seeds as they were constantly foraging for wild nuts, seeds, and fruits. Hemp seeds do not need any cooking or processing to make them digestible, which makes them a great snack for the hunter-gatherer on the go. The oldest archaeological record of hempseed being part of the human diet dates back 10,000 years, when it was discovered along with pottery shards attributed to the Jomon culture, an ancient early civilization of Japan (Okazaki et al., 2011).

Hempseed's nutritive value is even captured in the history of one of the world's great religions, Buddhism. A fable from the founding of Buddhism reported that Siddhartha Gautama, the founder of Buddhism, ate only a single cannabis seed per day for six years while seeking enlightenment.

Dating back as far as the seventh or eighth century BCE, Chinese authors wrote about the utility of hempseed, covering subjects including planting and harvesting times, nutrition, medicine, and derived products (Clarke and Merlin 2013, 201, 204). Practitioners of traditional Chinese medicine found frequent benefits in hempseed use, including treating urinary or blood flow problems, growth of muscle fiber, treating constipation, increasing mother's milk, and others (Clarke and Merlin 2013, 204). Hempseed was eaten throughout most of China's history and are still eaten in many parts of the country today. Hempseed oil may have been one of the earliest cooking oils during the Han Dynasty (200 CE), though it fell out of favor with many cooks because of its poor storage stability and the discovery of other organoleptically pleasing cooking oils. Hempseed oil did remain popular as a lamp oil, as it was purported to produce little smoke.

In the Mediterranean, lore around *Cannabis* seed as a psychoactive substance likely contributed to its spread as a food source. The use of hempseed and cannabis writ large by nomadic cultures of the Pontic Steppe was well documented by ancient Greeks like Hippocrates. Ancient Greeks frequently interacted with cultures that spread through Mesopotamia, Persia, Siberia and Mongolia—the Scythians, Phrygians, and Thracians. Some scholars credit these cultures as having an outsized role in the spread of *Cannabis* throughout southern Asia, the Middle East, and eastern Europe (Warf 2014, 420). Archaeological evidence of hempseed in burial tombs attributed to Phrygians and Scythians indicates that cannabis both had a practical and ceremonial value to those cultures (Sherratt 1995, 27).

Herodotus spread the lore of the Scythians and their love of hempseed by observing that "the Scythians would howl with joy for the vapor bath" (as quoted in Benet 1975, 40). The vapor bath, according to Herodotus, was a practice where the Scythians would throw hemp seeds on top of hot stones in a ceremonial container, dome the container with a piece of cloth, and inhale the vapors. This translation probably does not capture the subtext of what Herodotus was trying to say. Neither hemp nor marijuana seeds are psychoactive, so more than likely the Scythians were adding the whole flowers—seeds and all—to the stones to . . . get stoned.

Nevertheless, the practice of eating hempseed, often in the form of confections, for psychoactive effect became popular among the Greeks and eventually spread into the Roman Empire. In the *Handbook of Cannabis Therapeutics*, James Butrica observes how literary evidence suggests both the Greeks and Romans ate these confections during feasts and gatherings to discuss ideas and were well acquainted with their psychoactive effects (quoted in Clarke and Merlin 2013, 207). The hempseed cakes would promote euphoria and hilarity, indicating that they shared more in common with modern day marijuana edibles than as a means of nutrition. The Romans, notorious for being promiscuous eaters, no doubt ate hempseed for nutrition but there is little evidence to date suggesting that it became a staple food among wealthy Romans, who were famous for documenting their diet. Hemp seeds were even recovered from the ruins of Pompei after the eruption of Mount Vesuvius; however, it was not clear whether the seeds were being stored for planting or eating (Ciaraldi 2000, 91). But hempseed's presence in archaeological sites indicates widespread cultivation of hemp throughout Europe, meaning it more than likely was eaten as a peasant food. Traditional herbal remedies calling for hempseed as a medicine come up frequently in eastern European nations such as Russia, Ukraine, and Lithuania; no doubt these remedies have roots in folkloric traditions dating back to hemp's spread through the Roman Empire.

During the Middle Ages, hempseed grew became a good luck symbol in Europe. The French had a saying, "Avoir de la coire de pendu dans sa poche," meaning may you have hemp in your pocket . . . a colloquialism for good luck (Lee 2012, 20). From Ukraine to England, young women used to carry hempseed in their pockets to help them find love and a partner, and in certain Slavic countries the tradition of throwing hempseed on newlyweds for good fortune is still practiced (Booth 2003, 69). Festivals to bring about bountiful hemp crops were common in Europe, along with ceremonies to help it grow tall. Many German traditions included planting hemp seeds on the birthdays of notably tall saints, sowing hemp seeds by throwing them high in the air, and jumping over bonfires with the hopes that their leaps would encourage the crop to grow as tall as their courage.

Many of these historical traditions fuse the nutritive properties of hempseed with mysticism, healing, and reverence for the dead. These traditions likely endured in parts of Russia and eastern Europe from ancient Scythian traditions. Baltic nations like Lithuania make a hempseed soup called *semieniatka* to honor the dead believed to visit their families during the Christmas holidays (Benet 1975, 43). These traditions lost favor with many European cultures as they shifted away from hemp as a staple crop, but small pockets endure to this day.

HISTORY OF PSYCHOACTIVE USES OF CANNABIS

The utilization of hemp for fiber is one of the clearest through lines, spanning millennia across multiple civilizations. Not until the last one hundred years or so has hemp been banned in any major way around the globe and, even then, the ban was not very thorough. It does beg the question: if *Cannabis* was utilized for hemp fiber, were its flowers also utilized at the same time for their psychoactive potential? Were there cultures that used *Cannabis* only for fiber or only for drugs?

Certainly, there is a marked genetic difference between hemp cultivars and marijuana strains as well as between their associated cannabinoid profiles. The invention of ultra-low THC cultivars of hemp is a relatively new development in the history of cannabis agriculture. Historical "hemp" cultivars, while certainly not hitting the freakishly high THC levels typical of some current marijuana strains (20% by weight), generally still contained a few percent THC by weight, making them mildly psychoactive and, by extension, marijuana by our current definition scheme. Growing hemp in more temperate climates like Russia and northern Europe would suppress the expression of cannabinoids to a degree, but the historical record shows that humans did utilize cannabis for its psychoactive and medicinal effects.

There's further evidence in places such as India, where more powerfully psychoactive marijuana varieties took hold, that the plants were still processed for their fiber and seeds, similar to how we currently process industrial hemp. The point being is that from the dawn of history until the contemporary post-war period our knowledge of the specific mechanisms by which cannabis can exert its psychoactive effects was rudimentary at best. There is also an abundance of historical records showing societies knew of the medicinal and psychoactive power of cannabis and little evidence that past civilizations drew much of a distinction between hemp and marijuana. Were past civilizations more tolerant of psychoactive drugs than our current telling of history suggests?

A World Health Organization survey in 2017 concluded that roughly two hundred million people used cannabis for recreational or medical purposes in the previous year, roughly 2.6 percent of the global population (WHO 2018c, 4). Based on the size of this user base, cannabis is the third most popular recreational drug globally, only behind tobacco and alcohol.

While there are instances throughout history showing that humans theorized tobacco or alcohol could have medicinal properties, most medical practitioners now take a narrower view of tobacco and alcohol's chemical or pharmacological benefits. For example, alcohol has been, and still is, used as a powerful sanitizing agent. The usefulness of cannabis tracks similarly to tobacco and alcohol; however, due to the relative inaccessibility of cannabis to medical researchers over the last hundred years, researchers theorize myriad unidentified medical therapies derived from cannabis. That would put cannabis products more in line with classes of drugs like opium, where there are approved medical therapies and illicit recreational drugs.

The intersectionality of drugs used for both medical therapies and recreational entertainment highlights many interesting value systems in cultures throughout history. Themes such as clinical understandings of dependence and addiction; modalities of governance and power structures; thought, education (as it pertains to curiosity and exploration of the human condition), and reason; and human connection in a physical, metaphysical, and symbolic way—these themes help elucidate how society accepts the use of psychoactive substances, and they contour the role of drugs in society at large and in individual choices.

Before the formation of organized societies, psychoactive plants and fungi played an important role in spirituality and divination (Rudgley 1995, 25). Many scholars theorize that early modern humans, and perhaps some other hominids, sought out psychoactive substances for the means of transcending the circumstances of their existence and also to communicate with the realms of their ancestors and their gods. Some even theorize that the consumption of psychoactive substances not only played a crucial role in the formation of spiritual and religious experiences but may have even played a crucial role in the development of higher order brain functions and human intelligence (Schultes, Hofmann, and Rätsch 2001, 94).

Religion plays an integral role throughout the history of cannabis use. Its early history allows us to gain greater levels of appreciation for religious traditions, but as religious belief moved out of its early stages and coalesced around major religious traditions, cannabis fell out of favor. The predominant theory explaining the disavowal of cannabis use in a religious context purports that, eventually, organized religions sought greater levels of commitment to their faiths and sought greater control over their adherents' lives. Most religions espouse temperance and sobriety as tenets of faith, yet religious practitioners throughout history have a well-documented love affair with intoxicants. It makes some sense to think that certain religions would permit intoxicants that promote docility and conformity while banning drugs that promote nonconformity, even apostasy. Marijuana, a drug known for inducing unconventional ideas, was seen as a threat to maintaining cohesion between religious adherents and the religious institution (Robinson 1996, 78).

In early civilizations, shamans played a central role in cannabis consumption. Shamans communicated with the spirit world, played roles as early medical practitioners, and gave spiritual advice. More than likely, consuming cannabis would have been highly ceremonial and have taken place in contexts that had great significance to the consumer, such as marking significant events in his or her life. Unfortunately, we do not have any documentation of specific examples that reveal the spiritual significance of cannabis use to these early people; however, we can infer much from both traditional communities and the archaeological record.

One such example comes from an archaeological site in the Xinjiang province in northwest China. The site, referred to as the Yanghai Tombs, is attributed to the Gushi culture, a seminomadic horse-riding people. In one of the tombs, dating back about 2,700 years ago, a man was buried with lots of ceremonial objects, indicating he was very important to those who buried him. Many of the objects related to horse husbandry, but, along with a bow, arrows, wooden cups, and musical instruments, around one and three-quarter pounds (789 g) of marijuana flowers were buried with him. Considering the number of burial offerings in the grave and the positioning of the items buried with the body, scholars concluded that this man must have been a shaman (Russo et al. 2008, 4173). In fact, due to the extremely dry climate, the marijuana flowers were so well preserved that the authors were able to do a chemical analysis to see what the cannabinoid content of such an ancient sample might look like. The flowers contained high concentrations of oxidized derivatives of THC, indicating they were ceremonially placed to honor the shaman and that psychoactive cannabis was important to his life.

Early consumption of marijuana with the aid of a shaman more than likely involved taking high enough dosages to produce hallucinations. Marijuana is a hallucinogen and its effects are most pronounced at high dosage rates. Such experiences would give users feelings of transcendence and perhaps could even convince a user they were speaking to the departed as well as the divine (Russo et al. 2008, 4173).

The earliest recorded examples of recreational and medical uses of cannabis come from China. China has a long and complicated history with marijuana; no other civilization utilized the seed and fiber from the cannabis plant as much as China and yet the use of marijuana for its psychoactive effects did not have the same level of acceptance throughout Chinese history. Much of this attitude can be explained by the influence of Confucianism, a system of philosophical and moral teachings that guided Chinese civilization for millennia. Among the tenets of Confucianism is that of sobriety, which partially explains the aversion to marijuana use in modern and historical Chinese societies.

Chinese knowledge of the intoxicating powers of marijuana far predate the teachings of Confucius and can be seen in the Chinese language itself. The word for hemp in Chinese is *má*. According to Clarke and Merlin (2013, 218), má has two root meanings:

(1) "numerous or chaotic," derived from the nature of tangled hemp fibers, also providing the source of its use as a collective noun to identify bast fiber plants in general, and (2) "numbness or senselessness," resulting from the physiological properties of the seeded inflorescences, which were used in early medicinal infusions.

HISTORY OF MARIJUANA BEVERAGES

It is important to understand a few things about consuming marijuana. The first important thing is that cannabis in its raw, natural state is not psychoactive. We will discuss the biochemistry of cannabinoids, the chemical constituents that make marijuana psychoactive, in chapter 5, but here are a few basics. There are a few cannabinoids that have an intoxicating effect in humans, but the principal component in marijuana is delta-9 tetrahydrocannabinol, or THC. However, THC is not naturally produced by the cannabis plant, but rather it is a degradation product of delta-9 tetrahydrocannabinolic acid (THCA), an organic acid that is naturally synthesized by cannabis. To form THC, THCA must be decarboxylated—meaning it must be treated to remove a carboxyl group. The fastest way to decarboxylate THCA is to heat it, which with marijuana would normally involve either burning it or cooking it.

The second important piece of information is that THC is lipophilic ("fat loving") and therefore hydrophobic ("water hating"). If you were to simply boil marijuana in a tea, the THC would not fully dissolve into the water. You must use fat or a solvent to extract the THC—things like cooking oil or butter work well, as does ethanol, provided it is in a high enough concentration.

When understanding how ancient peoples would consume marijuana, keep those two things in the back on your mind. If an ancient human wished to burn marijuana, which nowadays we would just call smoking, how would this person go about doing so? Specialized papers for use in smoking, and even smoking pipes, are relatively recent inventions within the last 500–600 years. It is certainly possible that crude pipes were fashioned from wood or other materials on hand; however, there is little information available to us to confirm or refute this. From the information available we know that, if they wanted to smoke marijuana, more often than not a person would simply throw it on a fire or hot stones and inhale the fumes, or perhaps burn it as incense. Recall that Herodotus mentioned Scythian shamans employing such a practice while keeping a piece of cloth over their heads to trap the vapors. Smoking in this manner, however, is highly impractical, inefficient, and mostly done in a group setting. Instead, most ancient traditions of consuming marijuana involved eating or drinking it. Since this is a book about cannabis and beer, we will focus on historical examples of beverages.

A great example of an early marijuana beverage tradition comes from the Sredni Stog culture, a semi-nomadic group that lived in the fifth century BCE in what is now Ukraine. Archaeologists and historians credit the Sredni Stog with domesticating horses first, ostensibly through the aid of cordage fiber from hemp. In fact, the Sredni Stog culture could be an ancestral culture to the famously cannabis-loving Scythians, though that linkage has not been definitively proven. The archaeologist Andrew Sherratt found further evidence of the use of marijuana as an intoxicating beverage through residues in pottery in burial sites. The residues indicated that the Sredni Stog produced a beverage by heating marijuana leaves in butter and then infusing that into a beer or wine. Furthermore, the jugs found in the burial site had distinct cordage impressions around the exterior that appeared to be ornamental, not utilitarian. Sherratt theorized that the ornamentation was used to not only document the importance of the plant to the people, but also to advertise the contents of the jug and the nature of the intoxicants within (Sherratt 1997, 385).

The addition of cannabis seemed to be a way of additionally fortifying a beverage with nutritious protein, fat, and—depending on the THC content—extra intoxicating strength. Beers of the time lacked strength because of rudimentary malting techniques; adding cannabinoids would have helped make the drink more psychoactive. One thing that is impossible to know is exactly how psychoactive was the cannabis crop grown by the Sredni Stog. We know the Sredni Stog were using cannabis-derived cordage fibers for the domestication of horses and the production of tools and weapons. That the flowers would have yielded some cannabinoids is plausible, so it is certainly possible that the cannabis the Sredni Stog cultivated would have likely imparted noticeable intoxicating effects as well as providing added nutrition.

The Sredni Stog culture was a part of a vast lineage of Pontic Steppe peoples that lived from Europe to the western edge of the Himalayas. The Scythians were likely related to the Sredni Stog. These cultures, generally known for their horse-riding abilities and love of cannabis, likely accelerated the spread of *Cannabis* beyond much of its natural evolutionary climate. It was a well-adapted traveler. *Cannabis* grew exceptionally well in the fertile regions of Iran, the Middle East, and the foothills of the Himalayas in India. Marijuana thrives in the Indian climate, though it is unknown how it was introduced to the area. The best guess is that Paleolithic humans brought *Cannabis* to India, based on the fact that its pollen does not show up in the archaeobotanical record of India until 32,000 years ago (McPartland, Hegman, and Long 2019, 691). It is highly unlikely that *Cannabis* naturally spread to India from its home range of the Tibetan Plateau—there are no river systems that would naturally carry cannabis seeds to that area and the Himalayas would have prevented any dissemination through wind pollination or animal dispersion.

India is home to the best-documented cannabis beverage tradition, the intoxicating beverage called *bhang*. Bhang can still be found today; it is the most enduring marijuana beverage tradition in the world, perhaps being around as long as the Hindu faith, some 4,000 years old.

Bhang's terminology can be a bit confusing; it is technically a preparation of the psychoactive female flowers of marijuana, which are ground into a fine paste, often with spices. Though *bhang* refers to a specific preparation of cannabis, the Hindi word for "cannabis" is also *bhang*. To add to the confusion, the term *bhang* is a general reference to the intoxicating power within whatever edible form the marijuana takes. Bhang can be added to a wide variety of foods and a number of teas and lassi beverages. Russo describes a simple preparation of bhang tea: "Cannabis leaves are pounded; mixed with sugar and black pepper; blended with a little water; and added to milk, thin yoghurt (lassi), or milk tea. More complex formulations may include ground nuts, spices, and aromatic resin" (Russo, 2007, 622).

India has multiple words for describing popular forms of marijuana: bhang, the paste used in edible confections and beverages; *ganja*, female flowers and leaves for smoking; and *charas*, a concentrated form of ganja produced by rubbing the flowers and collecting the resin. Most people today know charas by its Arabic name: *hashish*.

Bhang continues to be a popular beverage consumed by many Indians, especially on religious holidays. Cannabis is strongly linked to the Hindu faith, being explicitly written about in foundational religious texts such as the Atharva Veda and the Bhagavad Gita. It is considered among "the five kingdoms of herbs . . . which release us from anxiety" (Abel 1980, 13) and to sharpen the memory and alleviate fatigue (Warf 2014, 420). Bhang is often consumed during holidays and festivals celebrating the god Shiva, who is considered the Lord of Bhang. According to legend, one day Shiva fell asleep underneath a large bush; when he awoke, he decided to eat a few of the leaves of the plant that shaded his rest. After a few moments, he felt energized and rejuvenated. From that moment on, bhang was Shiva's favorite plant. In certain parts of India, cannabis flowers can still be found as offerings in shrines to Shiva on holidays celebrating him.

Various adherents to Hinduism throughout history have followed ascetic traditions, eschewing any earthly pursuits and devoting their lives to their faith. Often, Hindu ascetics used bhang as a means to transcend their humble existence and aid in their spiritual enlightenment. Today, bhang can be purchased legally in government sponsored shops, which often serve more as tourist magnets than institutions of the community.

Another major theme in the history of the spread of *Cannabis* throughout the world comes from the interconnectedness of historical trade and warfare. The Indians, combined with Pontic Steppe cultures like the Scythians and the Aryans, began many of the marijuana traditions that endure to this day. These cultures invented how to concentrate marijuana into hashish, which made the psychoactive resins easier to transport and store. Hashish soon became the commonest form of marijuana in the ancient world, although the name, an Arabic word, did not become popularized until the eleventh century. Cannabis intoxication has been mentioned in historical writings ranging from the Greeks, Romans, Egyptians, Hebrews, and Gauls. By the fall of the Roman Empire, most of the Old World had some exposure to cannabis.

Historically, the last two places *Cannabis* spread was Africa and the Americas. *Cannabis* traveled to Africa through several different trade routes, likely spread from the north in Egypt and the Horn of Africa down the east coast of the continent. From there it spread inland, making its way to western Africa. Eventually, Arab migrants and African traders developed extensive trade networks where hashish was a highly prized crop. *Cannabis* traveled to the Americas via the transatlantic slave trade, brought by both enslaved Africans as well as European enslavers. *Cannabis* spread rapidly through South America and the Caribbean. The native peoples of South and Central America were well versed in the properties of psychoactive plants and cannabis became one of many used by them. Eventually, conquest, trade, and enslavement brought *Cannabis* north where it became established in the highlands of Mexico as well as the lowlands of the southern United States. It was also around this time that the more traditional hemp plants crossbred with traditional marijuana plants and produced some of the infamous *Cannabis* varieties known to the Americas today. As slavery became institutionalized, enslaved Africans brought with them their own cannabis seeds, many of which were stronger marijuana varieties.

By the time the European colonial period was in full swing, *Cannabis* had become established one way or another on every inhabited continent. Wherever *Cannabis* traveled, it found many cultures who enjoyed its psychoactive effects and versatility as a medicine.

HISTORICAL USES OF CANNABIS AS MEDICINE

Our understanding of cannabis as medicine dates back thousands of years; the distinction of cannabis as either a medicine or a recreational drug is a relatively recent construct. Human understanding of powerful psychoactive substances has frequently blurred the lines between medicine and recreation, so the context of how cannabis is used helps drive the distinction between the two. Cannabis can reduce inflammation and act as a pain reliever, and countless laborers through history have used it to help relieve the aches and pains of the day, not to mention lighten their mood after a hard day's work. Is that medicinal or recreational use? It's probably a bit of both. Likewise, alcohol use throughout history shares many similarities to cannabis use, yet few people today would argue that alcohol has much unique medicinal benefit. While alcohol can be valuable in some medical contexts, such as a wound sanitizer, the negative long-term impacts (dependency potential) and short-term impacts (poor judgment leading to fatal decisions; negative pharmacological interactions) make it very difficult and generally unwise to discuss alcohol as a therapeutically useful agent.

Understanding cannabis's usefulness as a therapeutic requires a level of nuance that isn't often expressed by its proponents. Clinical studies show that it has real benefits for human health and deserves a rightful place in pharmacology. That being said, much like with alcohol, there are many negative consequences of using cannabis (Voklow et al. 2014, 2219), including but not limited to: dependency and abuse disorders, impairment of memory, learning and brain development, and respiratory disorders. One obvious parallel to understanding the historical use of cannabis as a medicine is the history of opium. Opium has been a part of pharmacopoeias as long as cannabis and its use continues to this day in the treatment of severe pain. We also know that opium and its derivatives are some of the most powerfully addicting substances known to man and opium has one of the highest abuse potentials as a recreational intoxicant. Doctors through the centuries recognized opiates had a place in their toolkit, but also understood that prescribing it could lead to their patients becoming sick later through addiction. This tradeoff between the immediate demonstrative clinical benefits versus the long-term personal and societal disadvantages is crucial to understanding the history of all drugs, especially cannabis. Studies show that opiates have more potential for addiction and harm than cannabis (Bonnet et al. 2020, 4), that cannabis can be a suitable replacement to opiates for pain management as well as help wean users off opiates (Lucas 2017, 4), and that it is highly unlikely to overdose and die from cannabis (WHO 2018c, 38). Cannabis's medicinal properties are interesting when one considers different dosage rates. It has what pharmacologists call "biphasic effects," meaning it can trigger opposite effects between high and low dosage rates.

Historically, medical practitioners used cannabis to treat a wide variety of ailments that mostly stemmed from pain and inflammation in the body. As with most facets of cannabis history, the first culture to document cannabis-based medicines were the Chinese. Cannabis was the crucial ingredient in the first documented anesthetic used in surgery: called *má fèi sàn*—roughly translated to "cannabis boil powder"—this anesthetic allowed its creator, Hua Tuo, to dramatically increase his knowledge of surgical techniques and the origins of disease. In addition to má fèi sàn, Chinese doctors also prescribed má (i.e., cannabis) for hundreds of ailments, including gout, rheumatism, constipation, and pain (Lee 2012, 5). Emperor Shénnóng, the father of traditional Chinese medicine, wrote of the medical benefits of cannabis in the *Běncǎo Jīng* (*Pen Ts'ao Ching*), one of the earliest pharmacopoeias, and described cannabis as one of the "supreme elixirs of immortality."

The spread of *Cannabis* throughout the world was accompanied by the spread of knowledge about its medical benefits. Doctors and healers from India to the Mediterranean spoke of its virtues. The Greek physicians Galen and Dioscorides wrote of the many medical uses of cannabis, especially for treating pain and inflammation. The Romans documented hemp oil as an effective reliever of both menstrual cramps and birthing pains. Medieval Arab doctors widely prescribed it for pain throughout the Middle East and Africa. Cannabis-based medicines were so well-established in Arabic pharmacopoeias that even after the rise of Islam, which prohibits intoxication, many Muslim cultures permitted its use. Muslim doctors even tried to establish a dose of hashish that would numb pain but not cause intoxication so as to avoid religious prohibitions (Booth 2003, 49).

Many of these ancient cannabis-based remedies survived after the fall of the Roman Empire through folk medicinal traditions. Many European nations used hemp through the Middle Ages to treat headaches, earaches, and toothaches, and began to adopt cannabis-based medicines during the colonial period. Before its prohibition in the early twentieth century, cannabis was a key ingredient in many patented and over-the-counter medicines in America. Drug makers such as Eli Lilly, Parke Davis (now a subsidiary of Pfizer), and Squibb (now Bristol Myers Squibb) all produced medicines containing cannabis between the American Civil War and World War I.

Cannabis's power as a therapeutic medicine allowed for its legal reintroduction to the Western world following decades of global prohibition. Patented medicines such as Marinol® (dronabinol) helped relieve pain and nausea for cancer and AIDS patients following its introduction in the late 1980s, allowing for governments to slowly rethink their prohibitionist stances. For most of the previous century, there was a largely successful campaign to suppress the development and dissemination of information that would support the therapeutic use of cannabis. Viewing cannabis as a legitimate medicine created the permission structure to rethink our collective stance on its prohibition. Were it less efficacious, it likely would have been prohibited for further decades. As you will see, the modern history of cannabis cannot be separated from our understanding of its legal history, so we will continue to look at cannabis from that context.

Did You Know?

(To read more about these fun facts and others, please refer to the titles recommended at the beginning of this chapter and Suggested Readings on p. 251.)

- Did you know that Shakespeare may have smoked cannabis? An archaeological dig of Shakespeare's back garden unearthed several pipes containing residues of cannabis, tobacco, and cocaine.
- Some scholars theorize that the Old Testament references cannabis. Translations from Exodus show references for holy anointing oil, which some scholars say references cannabis oil. Messiah means "anointed one" in Aramaic, which would put Jesus in a very different context!
- It took 120,000 pounds of hemp to rig the USS *Constitution*, also known as "Old Ironsides."
- In the nineteenth century, you could order cannabis tinctures, cannabis cigarettes, and a candy called Gunjah Wallah Hashish Candy from a Sears and Roebuck catalogue.
- Hemp was so important to the British colonies in America that it was considered legal currency there for much of the eighteenth century.

2

CANNABIS LAW AND LEGAL HISTORY

To understand how we arrived at the current US regulatory framework, it's worth putting into context the ebb and flow of global acceptance of cannabis since the late 1100s. This review has a Eurocentric bent, mostly due to the dominance of European economic and political philosophies over the last five hundred years. This dominance also brings with it the legacy of slavery, racism, and white supremacy; governments of Europe frequently used the prohibition of cannabis as a cudgel to wield their power and oppress millions globally. Frankly, the legal history of cannabis outside of Europe seems to be relatively straightforward and comparatively rational. While prohibition of marijuana occurred in several Asian and Middle Eastern cultures, there was less of a concerted effort to eradicate its cultivation and use. At worst, cannabis users in Asian nations faced ostracization from their communities but were rarely imprisoned for consuming cannabis. European history, by contrast, displays multiple episodes of cannabis prohibition that ingrained the notion of consuming marijuana as a subversive act, all while the growing of hemp continued as part of maintaining the economic and military might of European states.

It is worth noting that the legal delineation between hemp and marijuana is a recent construct. Some scholars go so far to say that the separation of hemp and marijuana stems from the days of racist colonialist thought, where Europeans knew their hemp was the same as the hemp they encountered throughout Asia, Africa, and the Middle East, but since the citizens of those nations ate and smoked the flowers of their hemp for intoxicating and "lustful" purposes their hemp must somehow be different (Borougerdi 2014, 55). This paradigm, which arguably has continued through to this day, based the distinction between hemp and marijuana on whether a grower was harvesting the plant for fiber, seed, and inflorescences (i.e., harvesting for hemp) or if the grower was looking to harvest it only for the inflorescences (i.e., marijuana). Based on our current legal understanding, most of these historic plants would nowadays be classified as marijuana due to having a THC content greater than 0.3 percent by weight, even though those plants often were being utilized in a way that we recognize as hemp. In this review, the language will reflect our current established definition.

MIDDLE AGES TO THE TURN OF THE TWENTIETH CENTURY

Recall from the previous chapter that large parts of the Roman Empire were devoted toward the cultivation of hemp. After the fall of Rome, the cultivation of hemp continued unabated, although there is little documented history of the cultivation and use of hemp during this period. When hemp reemerges in the historical record in the twelfth century, it had a reputation as a folk medicine (Booth 2003, 71). Many superstitious uses of hemp arose across multiple European cultures, some of which continue to this day. Europeans broadly used hemp, both in folk remedies and in more pharmaceutical contexts, though there may have been little distinction at the time. Physicians documented hemp's efficacy in treating symptoms of pain, inflammation, depression, gout, labor pains, and many other ailments.

Eventually, marijuana became popular with members of the occult and in pagan religious traditions. Marijuana was a crucial ingredient to "witches brews" or black sacraments, ingested to commune with deities or to perform spells. Purportedly, these brews contained cannabis, opium, hemlock, and belladonna, which, according to accounts from the time, drove the drinker to ecstatic states, caused hunger, and was used as an aphrodisiac for orgies (Booth 2003, 72). These adherents to the occult drew the ire of the Roman Catholic Church. The

Church deemed pagan activities subversive and formally banned the use of marijuana in the fifteenth century at the outset of the Inquisition. This ban did not wipe out the use of cannabis throughout Europe, it merely drove the use and the understanding of it as an intoxicant underground. This is a common theme in the history of cannabis: governments or religions seek to subjugate and control a group, using the prohibition of cannabis as a tool to carry out their subjugation. This episode in history of cannabis suppression was deeply rooted in misogyny; women held much of the knowledge of folkloric medicines and provided medical care in many villages. Powerful organizations such as the Roman Catholic Church wanted to make sure the general population considered it as the one source of healing. The Church successfully demonized women by making the tools of their subversive power heretical.

While the Inquisitors of the Catholic Church killed heretics in Europe, Spanish conquistadors planted hemp throughout the New World to supply their colonial empire. The concurrent rise of capitalist and colonialist systems created a boom in the growth and distribution of cannabis throughout the world. The Russian Tsardom maintained a stranglehold on the growth and trade of industrial hemp throughout Europe, so the competing colonialist empires of Spain, England, France, and the Netherlands all sought to grow hemp throughout the world to lessen their dependence on Russian hemp. Colonialism and early capitalism required the exploitation of resources throughout the world and stable international shipments of these resources. Hemp comprised much of the essential raw material for rigging ships, so relying on any other competing empire for such a crucial material threatened the stability of colonial supply chains. As a result, each colonial empire sought to introduce hemp throughout their colonies, beginning in the early sixteenth century.

While hemp spread to the Americas, travelers returning from the Middle East and India reintroduced marijuana into mainstream European culture. Marijuana's reintroduction did not share a common origin or location in Europe, but rather it was a slow trickle of information based on trade, warfare, and travel by various Europeans. It is possible that pockets of marijuana use may have endured from the time of the Roman Empire, but written records show that many Europeans had little familiarity with marijuana during the age of colonialism. One of the earliest examples of this trickle of information came from the travelogue of Marco Polo, who wrote about a story he heard while traveling through thirteenth-century Iran. Marco Polo was told the tale of the Old Man of the Mountains, who trained and commanded a loyal fighting force that took a powerful drug before going into battle: hashish. The legend, according to Polo, stated that this group of mercenaries were known for their brutality in battle and their ravenous appetite for hashish. In fact, they named themselves after their favorite drug: the Hashshashin. We now know these people by a different name: the Assassins. While this story was a fabrication by Marco Polo, it did have some historical context. The story of the Hashshashin can be traced to a Persian sect of Shia Islam founded by a man named Hasan ibn-Sabah; however, most of the salacious bits have never been verified and are most likely slanderous fabrications. The story of the Hashshashin endures to this day and played a crucial role throughout history in reinforcing the notion that marijuana intoxication renders its user into fits of violence and insanity.

Variations on this trope that non-Europeans became wild, violent, and uncontrollable under the influence of marijuana became a predominant message told to white Europeans, which greatly influenced how they understood marijuana. Many scholars in Europe frequently commented on marijuana's ability to render humans into "savage" states, framing marijuana-using cultures as inherently inferior to European—white—society. This was just one of many pretexts Europeans used to justify their cruel and exploitative subjugation of millions. Make no mistake, Europeans had no abject aversion to intoxication; Europeans have long had an obsession with alcohol, a legacy that carries on to this day. This love of alcohol also brought with it shame and repressed desires, driven largely by the moral philosophies espoused by many Christian denominations and philosophers. Taken into context, it is easy to see how the general message formed that alcohol, especially drunk in moderation, can be a virtuous intoxicant, whereas drugs from exotic origins like marijuana would destroy any civilization it touches.

The ubiquity of alcohol in Europe may very well have kept hemp off the radar as a useful intoxicant. It may also have to do with a lack of knowledge of how to properly process and grow *Cannabis* for its psychoactive effects. European hemp varieties are relatively low in THC, but that is not to say they are devoid of it. Many *Cannabis* cultivars grown for fiber will reach one percent THC by weight in their flowers, a far cry from the 20–30 percent THC by weight found in some marijuana strains today. What we sometimes fail to appreciate is that, before recent decades, many marijuana plants were not much stronger than European hemp cultivars, many only reaching five to eight percent THC by weight in their flowers. In fact, as recently as 1995, the average THC concentration of marijuana seized by the US Drug Enforcement Agency was less than four percent by weight (ElSohly et al. 2016, 613). The processing technology required to make ancient cultivars strong was the creation of hashish. Hashish manufacturing requires the rubbing and sieving of marijuana flowers so that the glandular trichomes[1] separate from the vegetal matter. As the trichomes rupture, they get quite sticky and eventually form a crude paste. This paste can then be pressed into blocks for easy transportation and storage. Where a marijuana flower may only contain five percent THC by weight, hashish will contain upward of 30–40 percent THC by weight.

One of the main routes of introduction for marijuana into Europe during colonialism came through the transatlantic slave trade. European enslavers, recognizing that marijuana could be used to pacify enslaved people, permitted marijuana's growth on plantations. The Spanish, Portuguese, and French all introduced *Cannabis* into sugarcane plantations in the Caribbean, as well as Central and South America. Part of this introduction came from enslaved Africans, especially from western Africa. Many African societies used marijuana as an intoxicant; marijuana was originally introduced to African cultures through centuries of trade with various civilizations in the Middle East and India. The British slave trade between Africa and its Caribbean and American colonies provided an additional route for marijuana to enter the Americas, as did the British Empire's extensive network of indentured Indians throughout the empire. European slave holders in the New World knew that marijuana and alcohol helped maintain their plantations. Sugarcane plantations tolerated the use of marijuana and rum, knowing that enslaved people would tolerate their intolerable conditions by numbing themselves with intoxicants. Barney Warf (2014, 428) remarks that marijuana and rum were "intertwined in the cycle of work, debt, and poverty" that characterized life on sugar plantations and influenced global politics for generations. As slavery spread from Caribbean plantations to the North American continent, marijuana traveled with it. Slave owners initially permitted marijuana use in the South and that provided one of the routes of marijuana introduction during the colonial period in the US. This also created an association by the white, aristocratic slave owners between marijuana use and enslaved people, which came to inform later episodes in American history where white people would use marijuana prohibition as a means of oppressing Black people after the Civil War.

Cultural mixing through the slave trade introduced Europeans to many cultural traditions that endure to this day. Many people today associate the word ganja with the Rastafarian culture and religion in Jamaica, but *ganja*

---

[1] Glandular trichomes are the structures found in high concentrations on *Cannabis* inflorescences that produce cannabinoids and essential oils. We will discuss these in greater detail in chapters 3 and 4.

is a Sanskrit loan word from India. Additionally, the widespread practice of smoking marijuana came through contact with Native American civilizations, who smoked tobacco. With the exception of some parts of India, marijuana was rarely smoked prior to contact with Native American cultures but was instead mixed into foods, beverages, or tinctures. The introduction of smoking to Europe through contact with Native Americans and Indian cultures set off a smoking craze that dramatically increased Europe's demand for tobacco and marijuana.

In addition to the slave trade, Europeans also became familiar with marijuana from European physicians who traveled to the Middle East and India. These physicians often traveled with colonial armies and took an interest in the use of marijuana as a medicine by local doctors. Their findings were captured in medical journals of the time, discussing the many pharmacological benefits of marijuana they observed; they frequently spoke of its ability to relieve pain and to elevate moods. As these writings gained notoriety, a cottage industry formed to import hashish for further study. One of the most notable physicians to write about the medical benefits of marijuana was an Irish physician named William Brooke O'Shaughnessy. O'Shaughnessy served in the British East India Company during the 1830s and conducted a series of experiments to determine the efficacy and safety of marijuana for treating several maladies, including rabies, cholera, rheumatism, and epilepsy (Lee 2012, 24). He published his conclusions in 1842, noting that cannabis had few negative side effects and that it was effective as a pain reliever, muscle relaxant, and anticonvulsant. His findings generated tremendous interest throughout Europe and set off a wave of exploration into marijuana by doctors and patients alike.

O'Shaughnessy also worked to create an alcohol-based tincture, called Squires Extract, that exploded in popularity in pharmacies throughout Europe and the US. Encouraged by the medical efficacy of marijuana and curious about the mystical effects it could produce when taken in large doses, many creatives began experimenting with the drug. Marijuana-based tinctures were legal and accessible throughout Europe and the US during the nineteenth century, provided you were white and wealthy. Upscale social societies, such as the French Le Club des Hachichins (The Hashish-Eaters Club), were formed to hold parties where participants would have large doses of hashish edibles, drink wine, and discuss philosophy and mysticism. Notable writers of the time, including Charles-Pierre Baudelaire, Honoré Balzac, and Alexandre Dumas, participated in Le Club des Hachichins parties and frequently wrote about their experiences (Booth 2003, 84). Other social societies popped up throughout Europe and the US in the latter half of the nineteenth century to experiment with hashish and to write about its effects.

It was through these writers and thinkers that marijuana first became a countercultural drug. The nineteenth century is best described as a hedonistic century, where drugs and alcohol were ingested liberally. Alcohol and opiates were the drugs of choice for mainstream society. The marijuana-focused social societies helped expand marijuana's medical and recreational knowledge base through their writing and storytelling. Most of the writing tended to have exotic undertones to it, tying in Orientalist themes and fabrications about life in faraway places. This gave marijuana an air of inaccessibility, something taken by strange people in distant lands.

Around the turn of the twentieth century, marijuana use was at its peak throughout the US and Europe. It could be found in common over-the-counter drugs from reputable companies like Eli Lilly, it was featured in literature and plays, and had a similar standing to many other drugs in society. Marijuana was accepted as a vice and, while it wasn't exactly considered a part of upstanding society, few were ostracized for using it. At the time, it would have been farfetched to say that within a few decades marijuana would be banned and its users demonized.

HISTORY OF CANNABIS PROHIBITION IN THE US

The first instance of banning any drug in the US occurred in 1914 with the passage of the Harrison Narcotics Act. Beginning in the mid-1800s the US underwent a series of temperance movements that sought to eradicate the use of alcohol and other drugs from American society. These movements derived from genuine concern for the health and wellbeing of Americans—drugs and alcohol were abundant and easily accessible at the time. For example, in the 1830s Americans over the age of 15 drank on average seven gallons of alcohol per year, over three times the average amount consumed today (Rorabaugh 1991, 17). Drugs such as cocaine, opium, heroin, and marijuana were all available by mail order and prescription; their usage increased when opiate-addicted soldiers returned home from the Civil War. The Harrison Narcotics Act became the first US federal law to effectively ban the possession, use, importation, or sale of coca and opiate derivatives. Congress wrote the Harrison Narcotics Act to give itself the power to regulate and tax cocaine and opiates, but did not explicitly ban them from society. Congress wrote the law so that, in theory, one could obtain a license to dispense or use these drugs; however, in practice the government issued no licenses, effectively banning their use.

During this time, there were many concurrent forces in America working toward driving the prohibition of marijuana. The first was the movement to accurately label and improve the quality of the ingredients in food, beverages, and pharmaceuticals. At the turn of the twentieth century, manufacturers frequently sold counterfeit and adulterated consumer products and law enforcement had little regulatory recourse to prosecute them. Many patented medicines at the time contained opiates, marijuana, cocaine, and high-proof alcohol; until 1906 there was no federal legislation requiring drug manufacturers to label the ingredients that went into these medicines. The second driving force was the determined effort of several industries to ban the cultivation and sale of hemp products, most notably the cotton industry and the paper industry. Hemp cultivation had been in decline in the US in the latter half of the ninteenth century due to the increased adoption of technologies disruptive to traditional hemp markets: the cotton gin and the steam engine. The cotton gin had brought down the price of manufacturing cotton textiles to a point where hemp was no longer economically competitive in the textile industry. The steam engine had disrupted the maritime business's demand for hemp because steam allowed ships to sail without the aid of wind. As a result, ships no longer required the extensive rigging and sail cloths manufactured from hemp. As these technologies reached mass adoption throughout the nineteenth century, another cultural force concurrently began to push for the ban of marijuana: the temperance movement. While the temperance movement of the nineteenth and early twentieth centuries initially sought to eliminate alcohol from society, as public awareness grew about the dangers of other drugs, especially opiates, the temperance movement made prohibiting these drugs another of its goals.

The final—and primary—force driving the prohibition of marijuana was racism. At the start of the twentieth century, the largest demographic group in the US to frequently use marijuana were Mexican Americans, who were largely concentrated in the desert areas of the Southwest (Bonnie and Whitbread 1970, 992). The outbreak of the Mexican Revolution in 1910 forced hundreds of thousands of refugees to flee Mexico and head north into the US. Many white Americans resented this influx of Mexican immigrants and saw an opportunity to demonize them through their use of marijuana. Newspapers and yellow journalists further fueled this growing association of marijuana with Mexican immigrants by writing thinly veiled racist tropes about immigrants burglarizing houses, murdering (white) Americans, and seducing (white) children (Warf 2014, 429). These racist tropes would endure and spread throughout America for decades. These journalists seized on Marco Polo's story of the Assassins, written some 700 years before, to demonize African-American and Mexican-American communities and whip up a moral panic among white Americans:

> The more often the story of the Assassins was told, the more ludicrous it became. The image of the demented, knife-wielding, half-crazed hashish user running senseless through the streets, slashing at anyone unfortunate to cross his path, became part of the American nightmare of lawlessness. (Abel 1980, 224)

The classic example of this engineered moral panic can be found in the 1936 propaganda film, *Reefer Madness*, where a group of young, white students commit a litany of brutal crimes after smoking marijuana. To further reinforce racist stereotypes of marijuana users, when the white characters smoke marijuana, they begin to take on the caricatured, racist, stereotypical physical appearance of African Americans such as darkened skin, bigger lips, and bulging eyes. While today most people recognize the film for what is is—a racist, ridiculous piece of fear-mongering propaganda—at the time *Reefer Madness* was highly influential in convincing white Americans that marijuana was dangerous and even deadly.

Prohibitionists did not fully achieve their racist agenda until marijuana gained popularity with the emerging African-American middle class, led by the jazz musicians who represented a newfound level of African American influence on popular culture. Marijuana exploded in popularity along with the rise of jazz; many popular musicians, including Louis Armstrong, credited marijuana with fueling their creativity and talents. Marijuana smoking became synonymous with the enjoyment of jazz. Many jazz clubs allowed patrons to openly smoke marijuana in clubs and became de facto distribution hubs of marijuana in major cities throughout the US. The racist reaction to the rise in the cultural relevance and popularity of jazz was just one part of the oppressive and violent milieu that African Americans had to face. At a time when white racists murdered upward of 200 African-American community leaders and business owners in Tulsa, Oklahoma in 1921, politicians fueled by white racist ideology began seeking the criminalization of the activity associated with enjoying jazz: smoking marijuana. Racists looked for any opportunity to slow the rise of the Black middle class in America and found multiple ways to stymie African Americans seeking to achieve the American dream. Our history shows that the predominant form of racist violence took a more genteel form: criminalizing activities associated with communities of color and using the police and criminal justice system to terrorize and incarcerate members of these communities.

Following a decade of state and local ordinances, the possession or use of marijuana was officially banned nationwide when Congress passed the Marihuana Tax Act of 1937. The Marihuana Tax Act placed the regulation of marijuana under the Federal Bureau of Narcotics (the predecessor to the Drug Enforcement Agency) and placed a tax on the production, distribution, and use of marijuana products. Interestingly, the Marihuana Tax Act was the first time the term *marijuana* was codified in law. At the time, marijuana—often spelled *marihuana*—was a slang term for high-THC marijuana grown in Mexico that caught on in American popular culture. Some argue that the tax act of 1937 used the *marihuana* spelling to deliberately reinforce its association with Mexican Americans. The Marihuana Tax Act followed much of the same legal structure of the Harrison Narcotics Act of 1914, where the federal government would license, tax, and regulate the distribution of marijuana, especially because marijuana was still incorporated into several patented medicines. The Marihuana Tax Act quickly and effectively outlawed the legal use of marijuana in the US because the federal government would not issue any licenses to sell products containing marijuana.

In 1942, Congress and the United States Department of Agriculture (USDA) briefly allowed licenses under the Marihuana Tax Act for the cultivation of hemp fiber to aid the US military during World War II. During the war, the Japanese blocked all exports of hemp, sisal, and jute fibers from the Philippines, cutting off a crucial source of cordage fiber to the US. In order to make up for this, the USDA ran a promotional campaign called Hemp for Victory that encouraged farmers to grow hemp and process it for cordage fiber. The campaign was highly successful: the US went from producing less than 500 tons of hemp fiber in 1933 to 63,000 tons in 1943 (Robinson 1996, 162). After the Philippines was liberated from the Japanese in 1945, the USDA discontinued the Hemp for Victory program and all forms of *Cannabis sativa* went back to illegal status. The legacy of this program can still be found throughout the American Midwest to this day; hemp cultivated during the war escaped into the environment and still grows wild along creek beds and irrigation ditches throughout the region.

Congress increased the penalties for growing, possessing, or distributing marijuana in 1951 with the passage of the Boggs Act, which introduced the first mandatory sentences for drug-related convictions. After passage of the

Boggs Act, a first-time marijuana offender faced a minimum sentence of two to ten years in prison. This legislation was an early predecessor to the mandatory minimum sentencing laws of the 1980s and 1990s that were put in place, in theory, to dissuade citizens from using drugs. Many of these laws were particularly draconian: in some instances drug offenses could simultaneously be subject to federal, state, and local laws and an accused offender could be tried twice for the same crime (Schlosser 2003, 54). In practice, these laws destroyed communities and disproportionately imprisoned Black Americans and Latino Americans even though these communities used marijuana at similar rates to white communities. To learn more about this history and its continuing effects, I recommend you read Michelle Alexander's *The New Jim Crow: Mass Incarceration in the Age of Colorblindness* for an authoritative analysis.

The final piece of marijuana prohibition legislation, which remains in effect as of 2022, was the passage of the Controlled Substances Act of 1970. The Controlled Substances Act sought to clarify the federal government's position regarding the regulation, sale, and distribution of drugs. In it, the legislation provided different levels of scheduled substances indicating their medical value contrasted with their potential for abuse. Within this framework, the scheduling of a given substance gave guidance to doctors, researchers, and law enforcement in how to study these drugs and determine the best path to regulate and distribute them. The harshest scheduling was Schedule I, meaning this class of drugs had no demonstrated medical value and a high potential for abuse. The Controlled Substances Act became law despite multiple scientific inquiries into the health and safety of marijuana recommending that marijuana be decriminalized and regulated. In 1972, the Nixon administration, ignoring the findings of the Schafer Commission that recommended removing cannabis from the Controlled Substances Act and adopting other methods to discourage its use, instead decided to pursue policies that would later turn into the "War on Drugs" (a term later coined by the Reagan administration). One such policy included flying covert missions into Mexico to spray defoliants such as paraquat on marijuana fields to reduce supply (Craig 1980, 556). This policy failed to reduce the supply of marijuana in the US but did have two unintended consequences. The first was that marijuana farmers and smugglers chose to focus more attention on higher margin narcotic crops like poppy (heroin) and coca (cocaine), leading to the rise and funding of the violent drug cartels that are still with us today. The second was that marijuana farmers adopted new tactics to evade detection of their crops, developing indoor hydroponic cultivation techniques. These techniques dramatically increased the quality and potency of marijuana crops and expanded the genetic diversity of marijuana strains through advanced breeding. These growing techniques are now standard practice for the marijuana industry because the facilities can be secured and indoor operations can operate year-round.

After decades of prohibition, cannabis legalization started to gain support in the 1990s. California passed Proposition 215 in 1996, which became the Compassionate Use Act that legalized the use of marijuana for medical purposes when recommended by a doctor. The program became popular with the residents of California and eventually spread to neighboring states on the West Coast, including Alaska, Washington, and Oregon. At the same time, the Canadian government legalized the cultivation of hemp, starting a new cottage industry. The US government took note and legalized the importation of hempseed, seed oil, and fiber, although hemp remained illegal to grow in the US. As each of these programs gained popularity, cannabis-friendly local and state governments began rescinding some of the toughest marijuana laws within their jurisdictions. This included decriminalizing possession of marijuana up to a predetermined amount and recommending people arrested for marijuana crimes be placed into treatment rather than jail. Changes like this eventually led to state legalization. In 2012, Colorado and Washington passed marijuana legalization measures via ballot referendums. This marked the first time since the Franklin Roosevelt administration that a citizen could legally purchase and possess marijuana somewhere in the US.

Industrial hemp took a large step in its legalization journey in late 2014 when Congress added provisions to its omnibus farm bill, allowing for legal research of industrial hemp. When passed as the 2014 Agriculture Act, these provisions created a framework to allow states to legally cultivate industrial hemp, defined then and now as any *Cannabis sativa* plant containing no more than 0.3 percent THC by weight. The act allowed each state to come up with rules for its pilot research programs, which would then be approved and overseen by state departments of agriculture and the USDA. The goal of the legislation was intentionally broad: the pilot programs were to study market interest in hemp-derived products. Early adopters, such as Colorado, Montana, and Kentucky, all saw large-scale planting of industrial hemp. The largest consumer interest quickly arose around cannabidiol (CBD)[2] prod-

---

[2] Cannabidiol (CBD) is the primary cannabinoid produced by hemp cultivars. It is non-psychoactive, but has many interesting properties of interest to humans. We will discuss CBD in greater length in chapter 5.

ucts, which were marketed on the basis of unproven health benefits outside of CBD's clinically proven anti-epileptic properties. Many of these marketing claims came from varying levels of clinical studies that showed anti-inflammatory, anti-anxiety, antispasmodic, and analgesic properties.

The 2014 Agriculture Act did not fully legalize hemp, but rather set up a framework to better inform how to legalize it. A viability report issued by the USDA found that there was too much variability between states in the regulation and enforcement of hemp law, especially when trying to transport hemp across state lines. While some states provided clear and comprehensive legislation to regulate hemp, others either banned hemp or had unclear regulations about its legality. This brought up many interesting cases involving the shipment of industrial hemp products from a state where hemp was legal and regulated to another state where hemp products were explicitly prohibited. The conclusions reached by both the USDA and Congress showed that industrial hemp should be legalized at the federal level. Congress completed that work in 2018, with the passage of Agriculture Improvement Act, which legalized the cultivation and sale of industrial hemp products nationwide.

At the time of writing this book, marijuana is still not legal at the federal level. Since 2012, many states have joined Colorado and Washington in legalizing marijuana for recreational use and many others are introducing ballot measures to do so. My hope is that by the time you are reading this book the federal prohibition of cannabis will be a thing of the past.

Despite this hopeful note, it would be immoral to talk about the history of cannabis law in America without talking about its ongoing role in the systemic oppression and discrimination of racial minorities throughout the US. I'm sure many of you are reading this book because you are either looking for more information on cannabis or are interested in starting a cannabis business. We have seen how cannabis prohibition in 1930s America was instigated largely due to racial prejudice, but we must acknowledge the continuing toll this prohibition has exacted on communities in the US and elsewhere. There is a societal debt that must be repaid to those who suffered disproportionately for the policies put in place by our government. The US government has shamefully and deliberately targeted communities of color in the War on Drugs and exploited recreational drug use common to communities of all races as a pretext to disproportionately incarcerate members of communities that are not predominantly white. There are many great authors who write on this topic and I encourage you to explore their work. Authors like Kassandra Frederique, Jamelle Bouie, John Pfaff, Kojo

Koram, and Charles Blow (and many others) are all worth reading for their perspectives and prescriptions relating to mass incarceration and the War on Drugs. The Suggested Readings section has a list of some of my favorite articles and books on the subject (see p. 254).

Repeated studies have shown that racial groups throughout the US use marijuana at similar rates: roughly 12–14 percent of the population regardless of demographic. Despite similar use rates, members of the African American and Latino populations are on average 3.73 times more likely to be arrested and incarcerated than whites. In some jurisdictions, it's even worse than that, up to 15 times more in the worst case (ACLU 2013, 2). In 2010, the African American arrest rate for marijuana possession was 716 per 100,000 people, while the white American arrest rate was 192 per 100,000 people. Since the 1970s, the US government has spent $1 trillion fighting a drug war that has resulted in 40 million arrests. As a result of this drug war, the US now holds 25 percent of the world's prison population, despite only having 5 percent of the world's total population. In 2010 alone, there were 889,133 marijuana-related arrests. Put another way, police were arresting someone every 37 seconds for marijuana-related crimes. The American Civil Liberties Union found that, between 2001 and 2010, 88 percent of marijuana arrests were for simple possession. The near 890,000 marijuana-related arrests were 45 percent higher than the sum of all violent crime arrests that year, roughly 550,000 arrests. Despite the gargantuan levels of drug arrests, a World Health Organization study found that America's War on Drugs has had little if any effect on the use rates or availability of drugs (Degenhardt et al. 2008, 1053). Here are some of the effects the War on Drugs did have: funneled billions in revenue to violent drug traffickers, hampered the development of our knowledge and understanding of cannabis, discouraged those struggling with mental health and substance use disorders from seeking treatment, and prevented us from collectively developing consumer protections or quality standards for a products that millions consume annually. These effects demand reform and, thankfully, we are seeing progress.

To build on this early progress, we all must view reforming marijuana laws in the broader context of racial justice. There is a clear through line connecting the history of racism in the time of mass enslavement and colonialism to our current draconian drug policies, and that must be rectified. Justice will start when those incarcerated for non-violent marijuana crimes like possession are freed. Justice will continue when the criminal records for non-violent marijuana crimes are expunged. Justice will be served when tax dollars raised from the legal sale of marijuana are invested in the communities we destroyed. Justice will be served when we at last rip out every vestige of our racist marijuana laws that robbed so many communities of their humanity and dignity. Finally, we must guarantee equal access to entrepreneurs from communities of color to take part in the new cannabis economy, not just allow wealthy white people to buy access and opportunity.

CURRENT CANNABIS LAW IN THE US

An Overview of Current Marijuana Laws

Marijuana and marijuana extracts are currently illegal under federal law and classified as Schedule I Drugs under the Controlled Substances Act of 1970. The legal definition of marijuana encompasses the whole cannabis plant and any derivative substances obtained from the cannabis plant. Clarified after the passage of the 2018 Agriculture Improvement Act, any cannabis plant material containing more than 0.3 percent THC by weight is regulated as marijuana. The exceptions to marijuana's definition include:

> [T]he mature stalks of such plant, fiber produced from such stalks, oil or cake made from the seeds of such plant, any other compound, manufacture, salt, derivative, mixture, or preparation of such mature stalks (except the resin extracted therefrom), fiber, oil, or cake, or the sterilized seed of such plant which is incapable of germination. (21 U.S.C. § 802(16)(B)(ii))

These exceptions encompass the parts of the cannabis plant that do not contain or excrete psychoactive cannabinoids. Schedule I drugs or chemicals are defined by the federal government as drugs with no currently known medical use and with a high potential for abuse. Schedule I drugs include heroin, lysergic acid (LSD), peyote, methaqualone, marijuana, and ecstasy. Researchers can apply for a license with the Drug Enforcement Agency (DEA) to study these substances; trials are tightly controlled by federal regulators and few researchers obtain licenses to study marijuana due to its Schedule I classification.

While marijuana is illegal under federal law, numerous US states over the last thirty years have passed laws legalizing the use of marijuana for recreational and medical purposes. When federal and state law differs, generally federal law supersedes any state regulation unless Congress or the executive branch intercede to allow states to create laws that differ from federal regulations. A good example of this is the government's permission for dry counties throughout the US even though the sale and consumption of alcohol is legal. Dry counties exist because Congress included a carve-out in the Twenty-First Amendment that allows states regulatory authority over the importation of alcohol into their borders, effectively allowing the banning of alcohol in the state.

Marijuana's current regulatory framework exists where the executive branch has interceded to allow states to implement laws that differ from federal law. This occurs through official memoranda issued by a regulator or through discretionary enforcement—or lack thereof—by federal attorneys. The most well-known instance of this comes from the Cole Memorandum, issued in 2013 in response to the legalization of recreational marijuana in Colorado and Washington. The Cole Memorandum, issued by US Deputy Attorney General James Cole, instructed all US attorneys to not enforce federal marijuana laws in states that passed marijuana legalization measures and had established strong regulatory and enforcement mechanisms to control marijuana commerce in that state. It reinforced that the Department of Justice (DOJ) would prioritize the use of its resources toward controlling the illegal movement of drugs, including marijuana, by drug cartels and other violent drug trafficking organizations. Because it is the executive branch that passes these memoranda, enforcement discretion can change with a new presidential administration. During the Trump administration, then Attorney General Jeff Sessions rescinded the Cole Memorandum in 2018 and left marijuana enforcement to the discretion of the US attorney for each federal district. Currently, it is possible for US citizens engaging in the use or commerce of marijuana in a legal marijuana state to be arrested for violating federal marijuana law; however, it is highly unlikely that this will happen. Since legal medical and recreational marijuana markets create billions of dollars in revenue each year, it would be politically unpopular for a federal attorney to interfere with these markets.

The only act of Congress to clarify the federal government's position on state-regulated marijuana legalization comes from the Rohrabacher-Blumenauer Amendment. Representative Maurice Hinchey first introduced an early iteration of the Rohrabacher-Blumenauer Amendment into Congress in 2001 to prevent the DOJ from spending funds to interfere with the implementation of state medical marijuana programs. This amendment did not become law until 2014, when it was passed as part of an omnibus spending bill. Since this amendment was passed as part of a broader spending bill, it must be renewed each Congressional fiscal year to remain in effect. This amendment only applies to legal medical marijuana programs, which are now in 33 out of 50 states.

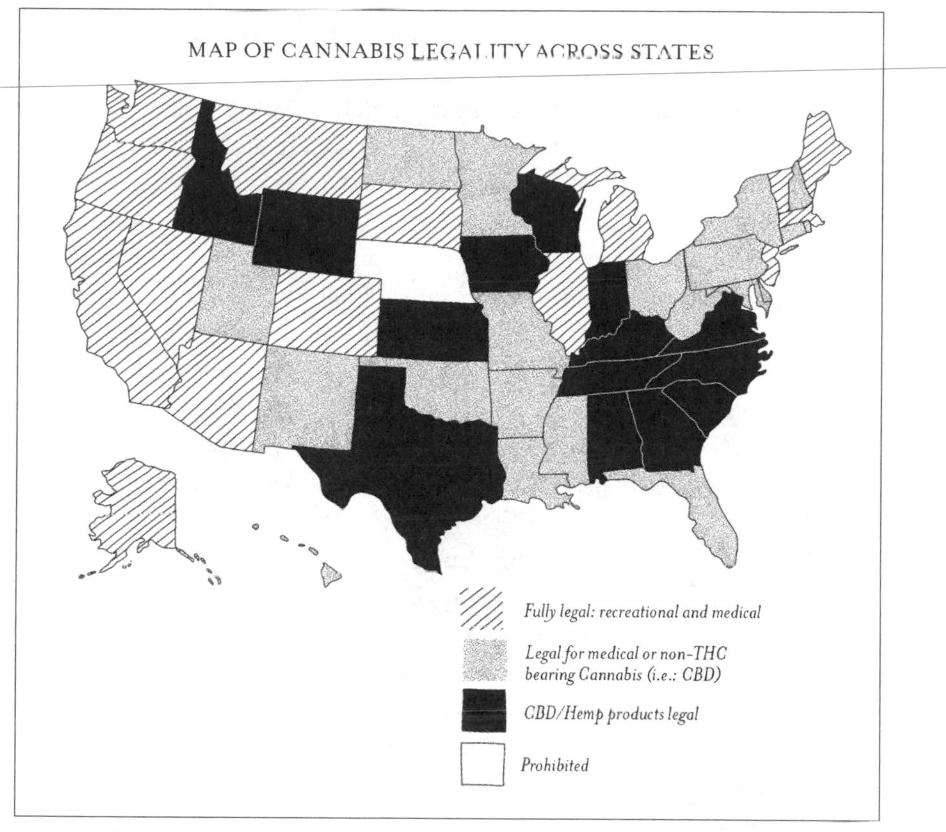

Figure 2.1. Currently, the states highlighted have approved marijuana for recreational and medical use ("fully legal"); for medical use or non-THC bearing cannabis products (e.g., CBD); or only for use of non-THC bearing cannabis (i.e., hemp cultivars and CBD products). Nebraska prohibits cannabis for any use. © *Hidden Powder Design*

Agencies that Regulate Marijuana

The DOJ and the DEA are responsible for enforcing federal drug policy. Most of this enforcement goes towards identifying and apprehending organizations involved in the interstate and international drug trade, such as drug cartels. While the DOJ and DEA could enforce federal marijuana law in states that allow for the sale of medical and/or recreational marijuana, they are not currently doing so at this time so long as businesses remain in compliance with state and local law.

The Food and Drug Administration (FDA) and the Alcohol and Tobacco Tax and Trade Bureau (TTB) do not allow marijuana in food or alcoholic beverage products. Despite states allowing the sale of marijuana-infused food and beverages, the FDA has declined to provide oversight or regulations for state marijuana markets.

At the state level, in states where such entities have been legalized, marijuana marketplaces are regulated by an assigned agency, usually the same agency that regulates alcoholic beverages. In most states, this agency writes all the compliance regulations for marijuana products that require accurate labeling, verifiable potency, allowable contaminant levels (e.g., toxins, pesticides, and heavy metals), and safe handling practices. Most of these regulations mirror federal regulations but currently the FDA is not actively involved in setting and maintaining these rules. In Colorado, for example, marijuana is regulated by the Marijuana Enforcement Division under the Department of Revenue and is a sister division to the Liquor and Tobacco Enforcement Division. For more information about marijuana compliance and how to successfully navigate the state marijuana regulations, please see Keith Villa's *Brewing with Cannabis: Using THC and CBD in Beer*, also published by Brewers Publications.

An Overview of Current Hemp Laws

Industrial hemp production became legal with the passage of the Agriculture Improvement Act of 2018. Any *Cannabis sativa* plant containing not more than 0.3 percent THC by weight qualifies as industrial hemp and any derivative product from that plant is legal. That is the easy part. Here is where things get tricky. For argument's sake, let us separate the plant into two buckets: the parts of the plant that do not contain cannabinoids and those that do. The first bucket contains the more "industrial" components of hemp such as fibers, hurds (the woody interior), and roots that can be turned into clothing, rope, paper, and construction materials. These components fit an easy regulatory scheme since they cannot be ingested for psychoactive or pharmacological effect and therefore do not need to be heavily scrutinized by regulators. The products made from them do have to comply with manufacturing or building codes, such as for making clothing or insulation, but these products do not require proactive testing requirements to verify their compliance with federal law. They are legal and can be manufactured with little oversight from the federal government.

The second bucket involves those parts of the plant—the inflorescences (the part that incorporates the flowers and related structures), leaves, and seeds—that can be turned into medicine, food, and nutritional supplements. Since these products will all be ingested by humans, the government provides additional levels of regulations to ensure the health, safety, and efficacy of these products. This material requires frequent testing throughout its growth and through the commercial supply chain to prove it complies with the 2018 Agriculture Improvement Act.

Proving compliance with the 2018 act ensures, among other things, that a farmer or manufacturer of industrial hemp does not inadvertently or deliberately produce marijuana. Testing should show the total composition of all cannabinoids present, including all THC content. The testing lab must give a detailed description of the testing methodology used and must maintain supporting evidence to defend the validity of the test results. Should tests show that a crop of industrial hemp is over 0.3 percent THC by weight, it must be destroyed in a process supervised by a state enforcement division or the USDA. Growers are given a fifteen-day window after confirming their crop is compliant to then harvest it, otherwise they must retest prior to harvesting.

The government regulates hempseed within two categories: as an agricultural product and a food product. As an agricultural product, the USDA regulates hempseed to guarantee crop quality. The USDA currently works with state seed certification programs, but it does not currently maintain a national seed certification program because hemp can overproduce THC in certain climates and underproduce it in others. Instead, the USDA regulates the importation of hempseed to mitigate the spread of any environmental pests. If hempseed is destined to be a food product it falls under the purview of the FDA. Generally, the FDA recognizes hempseed products such as dehulled hempseeds, hemp protein, and hempseed oil as GRAS, or "generally recognized as safe." This means that food manufacturers do not need special approval to add hempseed products in their formulations. When formulating alcoholic beverages, the TTB does require additional testing of hempseeds to guarantee they are free of any residual THC prior to use. A compliant test demonstrates the lowest levels of THC detected by the laboratory method, along with a test result indicating the level of THC detected or, more likely, that no THC was detected. With this analysis, the TTB should approve formulations of alcoholic beverages using hempseed, assuming the beer meets their other formulation criteria.

Cannabidiol (CBD) follows a complicated, and at this moment uncertain, regulatory scheme. The crux of the issue is that the federal regulatory agencies must determine whether CBD can be defined as a pharmaceutical drug. A helpful development came in mid-2020 when the DEA officially carved out CBD's definition from THC's definition, removing it from the Schedule I classification and rescheduling it as Schedule V. Before that time, any resinous component of the *Cannabis sativa* plant technically qualified as a Schedule I drug. The 2018 Agriculture Improvement Act recognizes that not all *Cannabis sativa* plants qualify as marijuana, so the resinous glands of industrial hemp do not fit the criteria of marijuana under the Controlled Substances Act. Any CBD extracted from legally defined marijuana still qualifies as a Schedule I drug, since its source material also contains more than 0.3 percent THC by weight.

This carve-out, however, means that CBD as a pure compound can be rescheduled and regulated. One CBD-based medication, Epidiolex, has FDA approval for the treatment of certain forms of epilepsy and was originally classed as Schedule V under the Controlled Substances Act. Schedule V drugs are defined as "drugs with lower potential for abuse than Schedule IV and consist of preparations containing limited quantities of certain narcotics" (https://www.dea.gov/drug-information/drug-scheduling). Schedule V is the lowest level of scheduled drugs and generally means these substances have a very low potential for abuse and are not considered habit forming. Since that time, the Drug Enforcement Administration officially descheduled Epidiolex and removed it from their purview under the Controlled Substances Act, meaning doctors and pharmacies that prescribe it no longer have to meet the requirements of state and federal drug monitoring programs. The sale and distribution of Epidiolex is still regulated by the FDA and it's only accessible with a prescription from a doctor and purchase from a licensed pharmacy.

So, what does this mean for all the CBD-containing products commercially available in the US? By the strictest definition of the law, they are currently illegal. In practice, it is more complicated than that. At the federal level, there is no law or regulation that expressly prohibits the sale of CBD products; likewise, there is no law that permits and regulates them. Three states, Idaho, Nebraska, and South Dakota, currently ban all hemp-based products, so they have better legal clarity than the rest of the nation. CBD products exist in a regulatory purgatory. Many manufacturers claim their CBD-containing products as dietary supplements and would be regulated under the Dietary Supplement Health and Education Act of 1994. Dietary supplements must meet minimum requirements of purity and must be labeled to describe every ingredient in the supplement. This claim becomes problematic, however, because the FDA has approved prescription drugs like Epidiolex that utilize cannabinoids like CBD so it cannot also approve the use of cannabinoids as dietary supplements. This is one of the many reasons the FDA has not officially ruled on the suitability of using CBD in food and beverage products; breaking this logjam may require an act of Congress.

In the absence of a ruling, companies making CBD products are working to proactively prove that their products comply with the Dietary Supplement Health and Education Act. The crucial component of this legislation is that the manufacturer cannot make *claims* about the benefits of taking a supplement.

Should a manufacturer make specific claims about the health or therapeutic benefits of taking a CBD-containing dietary supplement, such as saying their products can treat epilepsy, the FDA can shut down that business and confiscate its inventory. Therefore, you often see CBD-containing products in the form of tinctures, pills, and powders. These products fit neatly within the definition of a dietary supplement; they do not in themselves have nutritive value, so they cannot be misconstrued as a food or beverage. Producers must be very careful to not to claim direct effects of taking their product, otherwise they risk legal action from the FDA. Several companies have already received cease and desist letters from the FDA for making claims that their CBD products can treat medical conditions. Should a company continue making medical claims about its products, the FDA is within its rights to shut down the business and even prosecute the company for false and deceptive advertising practices.

The Agriculture Improvement Act of 2018 established that the US Secretary of Agriculture, along with the USDA, will set federal regulations and guidelines for the growth, manufacture, and distribution of industrial hemp and hemp-based products. Most of these regulations and guidelines pertain to issuing licenses to grow hemp and allowing certain fertilizers for growing hemp, but they also extend to the regulation of shipping and interstate commerce for hemp-based products. The USDA must also work with every state's governor and attorney general to draft a regulatory compliance plan for the cultivation and distribution of hemp and hemp products. The USDA must then approve each state's plan. There can be some variation in compliance parameters from state to state, so it is important to understand not only the federal guidelines pertaining to hemp but also the relevant state's compliance protocols. There are a few states that still prohibit the possession or distribution of hemp-products such as CBD, namely, Idaho, Nebraska, and South Dakota. This sets up a potential conflict between those states and the federal government over the rights of citizens to use federally legal but locally illegal products like CBD, as well as the states' right to confiscate and destroy hemp-containing products passing through their state in interstate commerce.

While the USDA regulates hemp growth and commerce, the FDA, along with the TTB, regulates the use of hemp products in manufacturing foods, beverages, nutritional supplements, and pharmaceuticals. The FDA ultimately decides the permissibility of any food and beverage ingredient, so the TTB defers to the FDA for any ingredient in the manufacture of alcoholic beverages. All alcoholic beverages that contain hemp products require a formula approval from the TTB.

Should a hemp grower or manufacturer deliberately run afoul of federal regulations, they will be referred to the state or federal branch of the DOJ—which contains the DEA—for investigation.

Labeling Practices for Hemp Beer

In addition to regulating the contents of beer, the TTB also regulates the way breweries can label and market beer. This regulatory authority covers everything from the physical packaging to advertisements to statements made by representatives of the brewery. There are several important labeling regulations that brewers must follow. The first encompasses what brewers can and cannot say on a label or through advertisements. In general, brewers are prohibited from making false or misleading statements about their products. Any beverage containing alcohol cannot be labeled or marketed as having a positive health effect in any way, nor can a brewer make statements that imply physical or psychological sensations resulting from consuming that beverage. Further, **any express or implicit statement that a beer can treat a disease or health condition is prohibited**. While a brewer may seek permission from the TTB to make health claims on its labels, those statements must be approved by the FDA. Brewers thinking of doing this should carefully consider the full ramifications of their ingredients and manufacturing process falling under FDA scrutiny—it is a level of regulation most brewers are probably not prepared to deal with. Finally, it is also illegal to direct consumers to a third-party source of information that provides biased or misleading information about the health benefits of alcoholic beverages or additives to alcoholic beverages. In short, directing your customers to "www.CBDBeerCuresCancer.com" is illegal. Directing consumers to independent, non-biased, third-party sources of information like the *New England Journal of Medicine* is permitted on a case-by-case basis and must be approved by a TTB official.

In the case of hemp or CBD, the legal guidelines for labeling and marketing follow the same principles as beverages containing alcohol. Manufacturers of hemp and CBD products cannot make claims that would imply any kind of outcome when consumed.

The TTB also has specific requirements regarding labeling and advertising hemp beers. The TTB prohibits the use of the term *hemp* on a beer label, except when it is part of a TTB-approved statement of composition, such as "Ale brewed with hemp seeds." Additionally, brewers may not use images or statements that imply or reference the presence of hemp, marijuana, or other controlled substances, nor can these images or statements in any way imply the beer will give a psychoactive effect from drinking it. These policies are subject to interpretation by the TTB and the officer responsible for label approvals.

Unfortunately, at time of publication of this book, the FDA has not provided further legal guidance clarifying the legality of adding CBD to food and beverage products generally and beer specifically. While there are many CBD-infused food and beverages out there, the manufacturers shoulder an enormous risk by producing and marketing these products. Yes, hemp is legal; however, in the absence of greater regulatory clarity, there is risk in adding it to food and beverages. If this is confusing, think of it this way: alcohol and tobacco are legal, but it's still illegal to add alcohol to a candy bar and or sell a tobacco-infused soda. Until the FDA provides a clear ruling, it is illegal to add CBD to food and beverages manufactured and sold in the US.

3

CANNABIS BIOLOGY, GROWTH, AND SUSTAINABILITY

CANNABIS SATIVA ECOLOGY AND TAXONOMY

Cannabis is remarkably well-suited for growing in a wide variety of environments. It can grow at altitudes of up to 8,000 feet (2,440 m), has a life cycle of only three to five months, and germinates within six days. It can grow at a rate of up to six inches (15 cm) a day, although 0.8–2.0 inches (2–5 cm) a day is more typical. Requiring very little water except during germination and early growth, cannabis will easily grow in poor, sandy soils; to realize its maximum potential, it prefers loamy soil. Conditions that mimic its evolutionary origins favor its growth performance: well-watered early in the year to simulate a flooding river, high nitrogen content to mimic animal waste, and disturbed, tilled soil all favor growth. As it is heliotropic, preferring direct sunlight, it does not thrive in shade and prefers open ground and high temperatures. Cannabis plants exhibit a trait called photoperiodism, meaning that it will grow while the summer days grow longer and will begin to reproduce as the days grow shorter. In areas around the equator where the change in sunlight is less dramatic and the weather remains temperate year-round, cannabis plants grow for several years, surviving multiple growing seasons. Accordingly, you can exploit this property by keeping a "mother" plant for cloning cuttings under artificial light that has constant daylight cycles.

Historically, some botanists thought the *Cannabis* genus contained three distinct species: *Cannabis sativa*, *C. indica*, and *C. ruderalis*. Carolus Linnaeus, the father of modern botany, first named *C. sativa* based on *sativum*, the Latin word for "cultivated." The French botanist Jean-Baptiste Lamarck suggested that European hemp varieties were physiologically different from Indian varieties and hence suggested they were different species; this is the origin of the classification *C. indica*.

Through modern genetic testing, we now know that the answer is a bit more complicated. Current research points to *Cannabis* consisting of just one species, *Cannabis sativa*, with identified subspecies and variants (Lynch et al. 2016, 358). Geneticists theorize that due to the domestication and spread of *Cannabis* it is unlikely that any cannabis plant today truly represents a wild population that evolved distinctly from other *Cannabis* populations. Considering how it can vary in size, density, growing characteristics, and chemical characteristics it is easy to see how early botanists were confused.

Understanding the genomic diversity within *C. sativa* is essential to understanding how it behaves and how to breed it. There are myriad implications for understanding the genotypic and phenotypic diversity of *C. sativa*; for instance, understanding the specific genes associated with producing cannabinoids could in the future allow growers to grow hemp that contains naught percent THC, removing many of the current challenges to growing legally compliant hemp. Similarly, many genetic traits associated with hemp, such as its high growth rate, could be exploited by a marijuana grower looking to boost yields or to create a more economic plant that could be harvested for both its potent THC flowers and its durable natural fibers.

There has been, and remains, some controversy over the taxonomic classification of existing varieties, cultivars, and strains of cannabis, but scientists agree all cannabis biotypes are of one species: *Cannabis sativa* L. Even though none are considered a separate species, for the purpose of this book, we will treat *C. sativa* as having three main biotypes with unique characteristics that warrant discussion. Remember, the variation of characteristics within *Cannabis* comes from a wide natural dispersal and evolution over millions of years, combined with significant human-directed breeding and domestication over thousands of years. The current agreed upon taxonomic classification would follow this order: *Cannabis* (genus) *sativa* (species) with two subtaxa, a subspecies (subsp.) such as *indica* and a variety (var.) such as *afghanica*. As such, the proper taxonomic term for the Afghani-domesticated plant colloquially termed "Indica" in the drug trade would be *C. sativa* subsp. *indica* var. *afghanica*.

The first biotype derives from characteristics that we colloquially understand as a "Sativa"-type plant. These plants are easily identifiable by their tall and gangly appearance and narrow leaf structure. We historically associate "Sativa" plants as European, Russian, and Chinese in origin; however, this is an oversimplification. Researchers often refer to these as narrow-leafed hemp (NLH) or narrow-leafed drug (NLD) biotypes to provide more precise language. From a taxonomic perspective, botanists classify NLH plants as *C. sativa* subsp. *sativa* var. *sativa* and NLD plants as *C. sativa* subsp. *indica* var. *indica* (McPartland 2018, 210). NLH or NLD biotypes can grow up to six meters tall (almost 20 feet) and generally feature wispy, open-faced branches and low-density flowers. NLH cultivars are prized for their strong, tall bast fibers and are most associated with the general term hemp. NLD biotypes are referred to as the "Sativa" drug types that colloquially give the user a more energetic and creative high.

The second biotype derives from characteristics we colloquially understand as "Indica"-type plants. These plants are squatter and bushier, with dense leaf and inflorescence structures. We associate "Indica" plants as historically originating from India, Pakistan, and Afghanistan; in the past they have been called by the names *afghanica*, *kafiristanica*, or *chinensis* (Small 2017, 461). Researchers often refer to these variants as broad-leafed hemp (BLH) and broad-leafed drug (BLD) types and, taxonomically, botanists classify these as *C. sativa* subsp. *indica* "hemp biotype" (East Asian hemp) and *C. sativa* subsp. *indica* var. *afghanica*, respectively (McPartland and Small 2020, 81). Many BLH and BLD cultivars originated from the Himalayas and the Hindu Kush region of Asia. In contrast with the narrow-leafed types, broad-leafed types grow only about a meter tall (just over three feet). BLH type cultivars are typically planted for cannabidiol (CBD) production due to their denser inflorescences compared with NLH cultivars. BLD cultivars are associated in the marijuana industry with "Indica" drug types that colloquially produce strong sedative effects. Many in the marijuana industry use the phrase "in da couch" to help customers remember "Indica" strains, claiming you will be glued to your couch after consuming them.

The final biotype is associated with a rare variant of *Cannabis sativa* called *ruderalis* or ruderal variants. It wasn't even recognized as *Cannabis* until 1924, when Russian botanists discovered it in Western Siberia. Botanists originally thought it was a unique species, calling it *C. ruderalis*. Since it was relatively isolated from other *Cannabis* biotypes, the theory at the time was it had unique characteristics adapted to its environment. Further research has shown that it is, in fact, a feral escapee of cultivated cannabis in the area. It is now known taxonomically as *C. sativa* subsp. *sativa* var. *spontanea*; or, if it contains sufficient quantities of THC to be classified as marijuana, it is *C. sativa* subsp. *indica* var. *kafiristanica* or, latterly, *C. sativa* subsp. *indica* var. *asperrima* (McPartland and Small 2020, 97). The ruderal variant displays unique growth characteristics, adapted to the Siberian environment of poor soils and short summers. Its main characteristic is it can autoflower, meaning it will automatically flower after about twenty to thirty days of growth. Compared to some of the more thermophilic *C. sativa* variants, *ruderalis* is imbued with the traits of many plants that survive in harsh environments; it must make the most out of a short summer growing season. Its autoflowering characteristic makes it unique among *Cannabis* biotypes—if the plant only has a short time to complete its life cycle it cannot waste precious time waiting to be pollinated. Ruderal variants are short and stocky plants, only growing to three quarters of a meter high (barely two and half feet) at full maturity, and have very little, if any, branching. In general, *ruderalis* produces low amounts of THC but high amounts of CBD, similar biochemically to cultivated European hemp. While apparently originating from the wild rivers of Siberia and parts of Russia, other *ruderalis* populations can be found growing wild across the northern latitudes, everywhere from the midwestern states of the US through many regions of Canada, and in many eastern European nations, including Ukraine, Belarus, Lithuania, and Latvia. Marijuana growers use the autoflowering characteristics of ruderal variants as a useful breeding tool for indoor growing operations. Hybrid strains with autoflowering genetics do not require complicated lighting systems to manipulate cannabis's photoperiodism, they will simply flower on their own.

CANNABIS GROWTH AND ANATOMY

Now that we've defined some of the genetic variations of *Cannabis*, we can explore its anatomy and how it grows. The main parts of the cannabis plant are the foliage, flowers, stalk, and roots. We will explore some of the parts of the cannabis plant necessary for its growth and reproduction here, mainly the flowers and foliage, then pick up how to transform these parts into useful products in chapter 4, which will include other parts of the plant as well.

The foliage of the cannabis plant provides metabolic energy. The leaves of the plant make *Cannabis* species readily identifiable. The leaves are characteristically palmate shaped and have between three and nine long and slender leaflets, depending on the cultivar. The leaves have a jagged appearance and range in coloration from bright lime green to dark green. They fan out from the plant, making it easily recognizable. There are two types of cannabis leaves: fan leaves and sugar leaves. Fan leaves grow up and down the plant and are the main centers of photosynthesis. Sugar leaves grow in close proximity to the flowers, ostensibly to provide an extra source of energy for the growing reproductive organs. Sugar leaves get their name because their secretory trichomes make it look as if sugar has spilled onto the leaf.

Flowers are the reproductive organs. Cannabis plants are primarily dioecious, meaning individual plants are either males or females and therefore produce only male or female flowers, respectively. *Cannabis* will occasionally display monoecious characteristics, meaning the plant will have both male and female reproductive organs. Flowering plants reproduce via sexual reproduction. The female flowers of *Cannabis* produce seeds during reproduction only with the help of the male flowers, which produce pollen to fertilize the females. Both sets of flowers grow in dense clusters at the top of the plant and at the nodes connecting branches with the main stem.

Cannabis Life Cycle

Generally, farmers adopt their growing practices to suit the needs of the female plants. Males most often are used for breeding purposes and are destroyed after they complete their reproductive role. This is a common characteristic in many cultivated plants, for example, hops are also dioecious with only the females commercially grown and harvested. In early growth, both male and female cannabis plants look very similar. This early growth period is called the vegetative state where the plant puts on much of its structural mass. Shortly after the summer solstice, the shortening days trigger the expression of sex characteristics and the reproductive cycle. Both male and female cannabis plants form flowers, growing first on the main stem, then spreading to the branch nodes on the rest of the plant. Once flowers begin to set, production of fan leaves (the broad-leafed foliage) slows dramatically as the plant begins putting the majority of its energy into sexual reproduction.

Cannabis follows these stages of growth: germination and emergence, vegetative growth, flowering, seed formation (reproductive growth), and senescence. Germination occurs in the spring after the last frost. Cannabis seeds can germinate as low as 32°F (0°C); they optimally germinate around 70°F, or 21°C (Mediavilla et al. 1998, 71). After about a week, the first growth will emerge from the soil and then the vegetative growth stage begins. The vegetative stage occurs over several weeks, depending on planting timing and the expanding day length of the photoperiod. During this stage, the plant will grow rapidly and add significant amounts of foliage. This is also the period when watering and soil nutrition is critical. Cannabis does require significant amounts of water during vegetative growth but can survive through drought conditions later in its life cycle. During the waning days of the vegetative stage, the plant will begin to flower at its nodes (the connective tissue that binds the leafy vegetation with the stem of the plant). During this period, the sex of the plant can be determined and, if necessary, the male plants can be destroyed.

As the photoperiod day length shortens, cannabis plants will switch their energy toward reproduction. Cannabis is a wind pollinated plant; however, it is possible for it to be pollinated by insects like bees. Both male and female plants will seek to maximize their chances of pollination by developing physical characteristics that maximize the chances of reproductive success. In the case of successful pollination, the female plant will produce a seed at its reproductive site in the flower and will move toward the senescence phase. If the plants do not successfully reproduce, they will continue to develop additional flowers and reproductive characteristics to try to successfully mate. They will do this until cold weather forces them into senescence. The senescence phase is the death phase of the plant, combined with shedding viable seeds from the flowers if they produced them. The plant will dry, wither, and eventually die; its seeds will carry on to the next growing season.

These growth stages represent how *Cannabis sativa* grows in the wild. Intentional cultivation of *C. sativa* uses these growth characteristics to optimally achieve the desired goal of the crop, so there can be some degree of variation in the life cycle stages. Before we talk about the different cultivation methods growers use for *C. sativa*, it's worth examining some of the differences between male and female cannabis plants to better understand why growers use different agricultural techniques.

Male Flowers of the Cannabis Plant

Male plants develop in ways that favor the dispersal of pollen. At maturity, males are typically slightly taller than females, but tend to be less dense and vigorous. Males usually have slimmer stems and fewer fan leaves. These characteristics show specialization for wind pollination; their gangly features allow the male flowers to shed pollen easily with even a slight breeze. The male flowers grow loosely and are relatively diffuse on the plant. As males reach greater levels of maturity, they become easier to identify and, if desired, culled from the female population.

Figure 3.1. A male *Cannabis* inflorescence and foliage. © *Hidden Powder Design*

Male flowers produce copious amounts of pollen. One study found that a single male cannabis flower can produce around 350,000 grains of pollen (Faegri et al. 1989). Male hemp plants are great for bees, which love the male flowers for how much pollen can be collected. Male hemp plants do not produce nectar in the same quantities that flowers of other species do, but their abundant pollen can be used as a vital food source in resource-deficient environments or late in the growing season when bees are storing food for the winter. In fact, some commercial honeys have trace amounts of hemp pollen in them. Bees do not visit female cannabis flowers, so they are not considered a vector for pollination in wild or cultivated cannabis plants. Since the use of pesticides is rare in growing hemp, researchers are studying the potential for hemp to be a major contributor to bolstering wild pollinator populations and in wild habitat restorations (O'Brien and Arathi 2019, 331).

Studies claimed that, under optimal conditions, cannabis pollen can travel upward of 180 miles, or 290 km (Clarke 1977, 87). Unsurprisingly, with the amount created by male plants and its ability to travel tremendous distances, cannabis pollen can be a major allergen. A study conducted by the National Institutes of Health found that during the peak production of hemp pollen in Nebraska, hemp contributed to 36 percent of the total airborne pollen count (Stokes et al. 2000, 238). At time of writing, there are no legal commercial hemp operations in Nebraska. All of the hemp contributing to the pollen count came from feral hemp populations that more than likely escaped from cultivated hemp plants dating from World War II.

Once the male plants shed their pollen, fulfilling their biological directive, their flowers drop and they die shortly after. Males do have some agronomic value after they have shed their pollen. Male cannabis plants used to be prized for the quality of their fiber, usually being harvested by hand back when labor was cheaper than today (Small 2017, 109). The fiber from males is generally much softer and makes wonderful clothing and textiles. Male cannabis plants do possess secretory glandular trichomes, albeit in much smaller quantities than females. (Trichomes are discussed in more detail below in relation to female plants.) As such, depending on the operation, male feedstock can be useful for extracting and purifying cannabinoids in either the industrial hemp or marijuana markets. Males do have secretory glandular trichomes on their foliage and stems, though very little in the flowers. In addition to extracting cannabinoids, males can also be commercially viable for creating distilled essential oil concentrates. We will go over the broad economic value of cannabis essential oil in further chapters—humans use cannabis essential oil for natural pesticides to cosmetics to my favorite use, which is flavoring beer.

The obvious function of male plants is for breeding. Breeding programs look for novel physical characteristics, superior agronomics, and fewer inputs for growing plants. New programs for breeding *C. sativa* are popping up in states with legalization and in some cases are merging with legacy clandestine breeding programs. Male plants provide half of the genetic makeup of every plant, thus it is important to understand its contribution in cannabis populations, even though most of the provided benefits exist in the background. Male rates of growth, pest resistance, drought tolerance, and general environmental resilience all must be understood to keep future populations healthy and provide maximum benefit to humanity.

Female Flowers of the Cannabis Plant

What we typically associate with *C. sativa* comes from female inflorescences and foliage. Female cannabis flowers garner much of the attention and for good reason. Female flowers produce fertilized seeds for reproduction and their glandular trichomes secrete resin containing the highest concentrations of aroma compounds and psychoactive cannabinoids. Female cannabis flowers

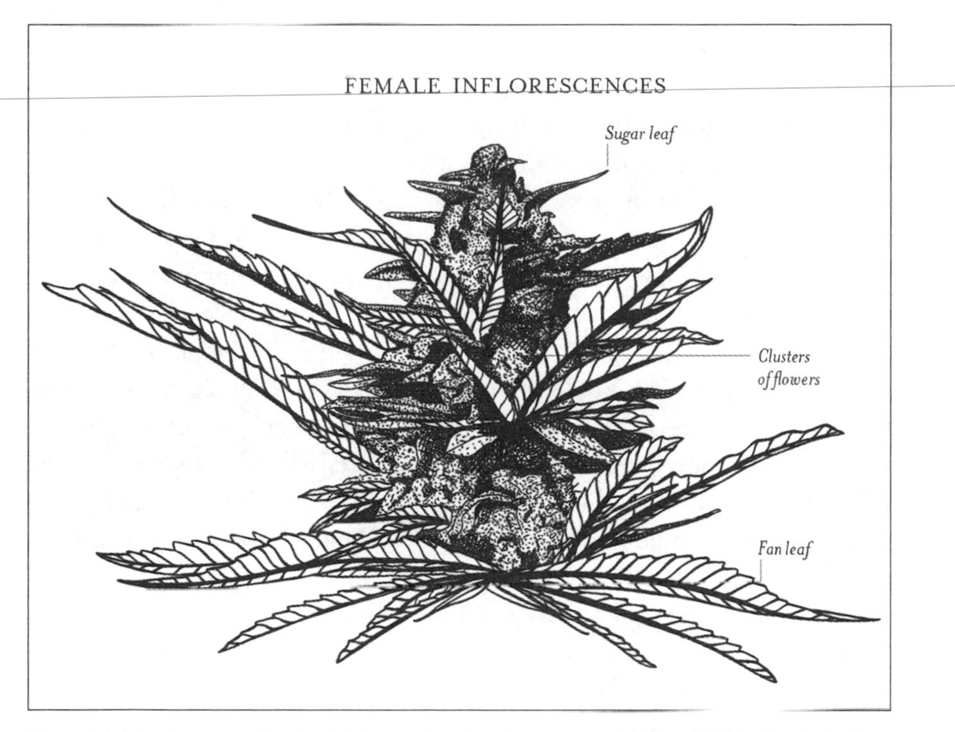

Figure 3.2. The primary constituents of the female *Cannabis* infloresence and foliage. © *Hidden Powder Design*

are technically **inflorescences**, referring to a group or cluster of flowers or the entire branching system of flowers. Terms such as "bud," "nugs," or "flower" are commonly used in marijuana jargon to refer to raw smokable cannabis material that contains high quantities of inflorescences. I will do my best to avoid these jargon terms throughout the book; however, if you see these terms, know they are in reference to the female reproductive material of the cannabis plant.

Female inflorescences grow in spicate cymes, that is, spiky clusters of petals. They are generally more clustered and denser than male flowers. Female flowers grow to have characteristics designed for the capture of male pollen and contain an ovary that has two stigmatic branches that grow out of it to catch wind-blown pollen. On the branches there are many papillae, tiny protuberances that act as pollen receivers. Like males producing a lot of pollen, females produce a lot of papillae to increase the odds of capturing fertilizing pollen. Unfertilized female flowers can grow very long stigmas in response to prolonged periods without exposure to pollen. This can be demonstrated in the marijuana industry by the many marijuana strains with very pronounced stigmas that also give the flower a distinctive color. In marijuana jargon, the stigmas are often referred to as "hairs" although they have no biological similarity to hair.

The ovary does not exist out in the open. The flower is wrapped in a **perigonal bract**. The perigonal bract plays a protective function for the ovary and, if the plant reproduces, it envelopes the growing seed. The surface of the perigonal bract is covered in glandular trichomes that secrete myriad substances that play a critical protective function for the plant. Many of these substances will be covered in subsequent chapters. The glandular trichomes are also where the plant synthesizes cannabinoids such as THC and CBD. There's a common misconception that the flowers of cannabis get humans high. This is not true. If you were to buy pure flowers (i.e., not the full inflorescences) of the cannabis plant, which are absent of trichomes, you would be disappointed and possibly slightly ill—some plant stigmas can be toxic to humans. It has been theorized, but not proven, that the stigmas of cannabis plants could have adverse health effects in humans (Naraine and Small 2016, 340). The glandular trichomes secrete all the cannabinoids and flavor compounds that humans have found so useful and irresistible; these cannabinoids are the primary compounds of interest to humans and so, of course, we spend our time and energy studying them.

Trichomes come in multiple forms and provide myriad functions during the life cycle of a cannabis plant. The evolutionary purpose of trichomes is to provide a primary defense mechanism to ward off predators and disease. It's no accident that the highest concentration of trichomes can be found surrounding the reproductive material of the plant; trichomes were an evolutionary advantage to help ensure the plant's genes carry on to the next year. Trichomes can be found in highest abundance on the perigonal bracts surrounding the pistils of the female inflorescences. They are also found in lower concentrations on any non-reproductive plant matter surrounding the inflorescences, including the leaves and stems. The stigmas do not themselves possess secretory trichomes, despite common misconceptions, but trichomes often break off and get stuck on the stigmas.

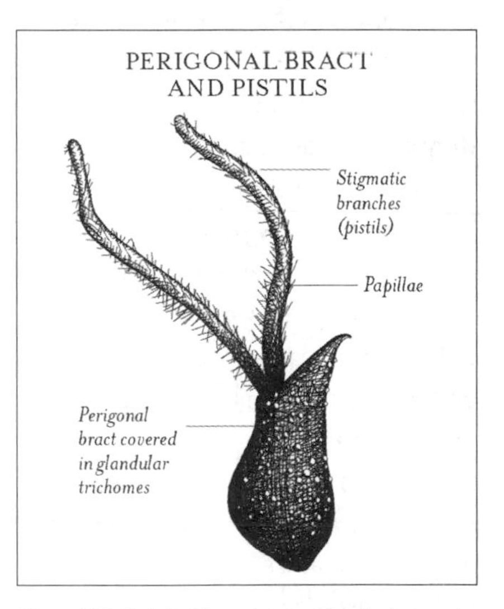

Figure 3.3. Detail of the anatomy of a single female *Cannabis* inflorescence. © *Hidden Powder Design*

There are two main classes of trichomes: unicellular and multicellular. The unicellular trichomes act as a primary physical defense mechanism for the plant. They have a hair-like look to them and are most often found on the foliage, flowers, and the branch stems and nodes of the plant. Unicellular trichomes come in two forms: cystolithic and simple. Both act as a mechanical defense against herbivorous animals and insects through different mechanisms. A cystolithic trichome contains calcium carbonate at the tip of its structure and calcium oxalate at its base. Both substances are quite bitter and unpalatable to herbivores; in fact, it is a fairly common antiherbivore adaptation in the plant kingdom (Small 2017, 87). In addition to their unpalatability, the calcium minerals make the plant hard to chew, so even if an animal decides the bitter taste is tolerable it still must chew rough, mineralized leaves. Simple trichomes act as a physical barrier, mainly for insects and fungi trying to infest the plant. Simple trichomes grow in dense, spiky clusters that look like microscopic fortification measures. Additionally, simple trichomes provide natural insulation for the cannabis plant. *Cannabis* evolved to tolerate arid climates and usually grows best when left open to the elements. Simple trichomes provide a natural mechanism to insulate the plant from desiccating winds and scorching sunlight.

Multicellular trichomes are far more complex. Multicellular trichomes are often referred to as **glandular trichomes** (or secretory trichomes), meaning that they excrete useful compounds into a cavity in the gland. In cannabis plants there are three types of multicellular trichomes: bulbous, capitate sessile, and capitate stalked. Of the three, only bulbous trichomes do not produce cannabinoids. Some researchers suggest that bulbous trichomes are another defense

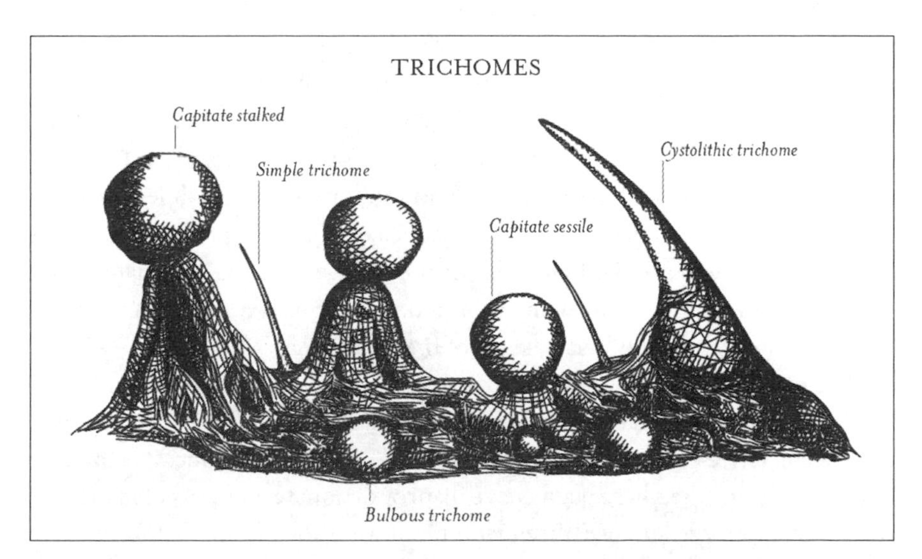

Figure 3.4. The different types of *Cannabis* trichomes. © *Hidden Powder Design*

mechanism against insect predation. When an insect steps across a bulbous trichome it will rupture the gland's cavity, ensnaring the insect in its resin and releasing signaling compounds to the plant to put its energy into additional defenses. The capitate trichomes, both sessile and stalked, produce the cannabinoids, essential oils, alkaloids, and other bioactive compounds that make *Cannabis* infamous.

As the growing year drags on and a female flower has not been pollinated, the plant will continue to grow more flowers. This in turn means it will grow more stigmas and more perigonal bracts to protect them; with more perigonal bracts, the flower is covered in more glandular trichomes. This process will continue until the weather no longer cooperates. Conceptually, this is how cannabis growers manipulate the plant into maximizing its trichome—thus, cannabinoid—content. All agricultural techniques stem from controlling reproduction between male and female plants. Now we will explore some of the main techniques used by cannabis growers to produce the main products derived from the plant: cannabinoids, terpenes and terpenoids, fiber, and seed.

METHODS OF CANNABIS PRODUCTION

Sinsemilla Production

Sinsemilla production from cannabis involves a variety of farming techniques aimed at growing only female plants and suppressing the growth of male plants, sometimes via extreme measures. A Spanish word, *sinsemilla* means "without seed" and denotes techniques originally developed by Mexican marijuana growers. These techniques became popular with indoor growers of marijuana in the 1980s as the US government began an active campaign to destroy outdoor marijuana farms in Mexico, often with the chemical paraquat (Warf 2014, 431). Domestic growers, seeking to evade detection from law enforcement, moved marijuana production into greenhouses and indoor growing facilities. Growers developed techniques that expanded on the idea of sinsemilla production, which are now considered standard growing practices for producing marijuana. Back before modern marijuana breeding techniques, sinsemilla was a common term used to denote the quality and potency of the marijuana.

The idea is simple: without fertilization of male pollen, female plants will continue to produce greater amounts of inflorescences—and their corresponding psychoactive resin—in their physiological quest to reproduce. At the end of the growing cycle, these plants have disproportionate amounts of glandular trichomes compared to a fertilized wild plant, thus increasing their economic value in recreational and medical marijuana markets. With the rise in demand for CBD products, growers of industrial hemp have increasingly adopted sinsemilla production techniques to boost the levels of CBD in their crops.

There are several techniques that can be used to prevent seed production. The most common technique is vegetative propagation of female plants, combined with the strict exclusion of male plants and pollen. This maintains the genetic integrity of the crop and guarantees that the clones will not be male. Growers and breeders will often select a particularly hardy plant that they will then designate as the "mother" to the crop. The mother will often be kept in a highly controlled environment where its watering, nutrition, and daylight exposure are artificially regulated. Since most cannabis plants grow and reproduce based off of the amount of daylight throughout the summer, one can keep a mother plant under extended hours of artificial light so the plant thinks it is still in growth mode. Theoretically, a mother can be kept in this state for years, so long as it doesn't get sick or is attacked by insects. The mother plant is then periodically trimmed to produce clones. These cuttings are dipped in a solution containing rooting hormones that will trick them into forming roots, at which point they can be transplanted into soil and will grow into brand-new plants. This process can be repeated thousands of times to produce whole crops in both indoor and outdoor farms.

Many crops undergo vegetative propagation, though not all crops use the cloning techniques that are popular with cannabis growers. Hop plants are also propagated, but by their roots rather than their foliage. Hop plants must also be grown in an asexual environment; hop growers will walk through their yards eliminating any males that would otherwise diminish the aromas of their crop. Cloning allows for a uniform genetic makeup and allows for far greater levels of control of the end cannabinoid and terpene profiles.

The sinsemilla technique is not used for the production of industrial hemp fiber or seed, mostly because the process is highly labor intensive and there is less need for genetic and sexual uniformity in those operations. The high labor inputs are not offset by the relatively modest per acre revenue that industrial hemp seed and fiber produce.

Seed Production

Cultivars used to produce seeds (and, generally, fiber), utilize traditional growing techniques that allow for sexual reproduction between male and female plants. Typically, these operations grow hemp for the purpose of harvesting seeds to be eaten or turned into hempseed oil. In certain breeding applications, cannabis plants are bred sexually for the purposes of creating new chemotypes (see p. 110) or to breed other new characteristics into the line.

When allowing natural reproduction between plants on a cannabis farm, it is important to ensure not only that the farm has the intended cultivar that will reproduce, but also that there are no other farms following the same practices. Male cannabis plants are prolific pollen producers and it is generally recommended to create an exclusion zone of five kilometers (just over 3 miles) when growing an all-female crop outdoors (Small and Antle 2003, 37). When growing for seed, normally most of the male plants are culled from the farm once they have reached enough maturity to identify plants by sex. Successful pollination of the crop requires few males—one or two male plants can pollinate surrounding females within 25 square meters, or about 270 square feet (Small 2017, 61).

Some farmers select for cultivars that display a high propensity for hermaphrodism, where a plant has both male and female reproductive parts and can self-pollinate. Some modern fiber cultivars, referred to as monoecious cultivars, will have both male and female flowers on the same plant. Wild monoecious *Cannabis sativa* is atypical but not impossible; more typically, hermaphrodism in cannabis is a trait that is bred into commercially managed varieties, eliminating the need for culling males from the field and leading to a more uniform crop. A downside of growing these cultivars, especially in successive generations, is they tend to display characteristics of inbreeding, such as stunted growth (Small 2017, 59). Monoecious crops are of the hemp biotype and are used as a "dual-purpose crop" where the plant can be harvested for both its seeds and fiber, generating extra profit per acre for the farm. Since the production of seeds suppresses the overall output of glandular trichomes, it is rare that a cannabis farm can harvest these crops for all three valuable outputs. Breeding efforts have sought to boost glandular trichome output for dual-purpose hemp crops, but little commercial success in these efforts has been reported. Theoretically, advances in processing technologies could allow for trichomes to be captured without sacrificing the quality of seed and fiber, but these technologies have not been developed to date.

Production from Feminized Seeds

Another popular growing technique is growing from certified feminized seeds. This has become increasingly popular with industrial hemp farmers who are growing crops to produce CBD and, to a lesser extent, terpenes. This technique captures the advantages of utilizing large-scale planting equipment with the guarantee that a crop is uniformly female. With industrial hemp farms growing upward of 500–1,000 acres per season, the labor requirements of sinsemilla production erode the profitability of the crop. Clones from sinsemilla production must be hand cut and often hand planted to ensure success. New technologies are helping with mechanized planting of clones but those technologies are rare in commercial cannabis farms.

Conceptually, producing feminized seeds requires techniques that trick female plants into producing seeds that will be exclusively female. The process requires adding chemicals to a female plant still in vegetative growth so that it begins to grow "male" flowers. The plant looks similar to a hermaphroditic cultivar but, genetically speaking, the plant is still 100% female. When the pollen hits the usual female flowers, it will produce seeds that only have female genetics, thus ensuring the seeds will only produce female plants. Additional treatments, such as subjecting seeds to ultraviolet light or irradiating them with gamma rays, have shown to increase the rates of female genetic expression without resorting to techniques to produce exclusively feminized seeds, but they do not produce exclusively female crops (Nigam, Varkey, and Reuben 1981, 390). Manipulating environmental factors such as nutrition, soil composition, temperature, light intensity, and plant mutilation can influence the expression of sex to a certain degree but cannot be guaranteed to produce one sex or the other (Small 2017, 60).

The specific treatments for producing feminized seeds involve hormonal and chemical treatment and, depending on the application, can add the opposite sex's flowers to either a male or female plant. For instance, treatment with ethylene gas, which is naturally synthesized by plants and functions as a hormone for regulating growth and development, will induce the formation of female flowers on a male plant. Conversely, techniques to suppress the formation of ethylene gas and its biosynthetic derivatives, such as treatment with silver thiosulphate and silver nitrate, can induce the formation of male flowers on female plants (Mohan Ram and Sett 1982, 369).

These techniques are highly effective so long as they are applied under controlled conditions. However, even under controlled conditions there are still occasional hermaphrodites and those plants can pass on genes for hermaphrodism. In 2019, there were many anecdotal reports of hemp farmers paying top dollar for feminized seeds only to find that upward of 50 percent of their crop was male. Those hemp operations were all growing for high CBD content; when half of their seeds produced little to no CBD and the other half produced dramatically reduced levels of CBD, it led to large losses for those farms. To guarantee true fidelity when maintaining genetically female plants, cloning techniques must be employed.

SUSTAINABILITY

Sustainability is an important topic in the hemp world, and it is one of the main reasons driving hemp's resurrection from agricultural obscurity. The potential to reap multiple products from a single harvest makes hemp highly desirable from a sustainability perspective. Sustainable agricultural operations seek to understand the level of resources required to grow a crop weighed against how many products can be made from growing that crop. An ideal sustainable crop balances the needs of the market with the health of the environment. Environmental health involves the direct effects of agriculture on the cropland itself as well as the effects on any land that agricultural may inadvertently alter through farming practices, including direct effects (e.g., runoff) and non-direct effects (e.g., greenhouse gas emissions from fertilizer production).

Right now, industrial hemp, when grown under strong environmental regulations, is considered one of the more sustainable crops compared to other major crops (Montford and Small 1999, 61). This is largely due to the fact that most of the hemp plant is harvested and utilized, it requires minimal bio-cides to ward of predation and disease, the products produced from the crop require minimal energy to process, and the end products are either recyclable or biodegradable.

Marijuana, on the other hand, is one of the least sustainable crops. It doesn't have to be this way but the way marijuana is grown currently tends to leave an outsized environmental impact, which should be addressed. Marijuana is often grown indoors or in semiclandestine environments that employ large amounts of chemical fertilizers. Clandestine black-market operators often grow marijuana on remote public land. They set up camp where few can find them and hike in all of the supplies they need to grow a crop. Since they lack incentives to maintain the environment they are grow-ing in, these operators usually just abandon their waste when they harvest the crop and move on to another location. There are plenty of instances where illegal marijuana operations have dumped fertilizer, biological waste, and garbage into habitats; this poisons native animals and the waste runoff poisons water systems, causing aquatic dead zones.

We will look at a few factors that drive hemp's environmental sustainabil-ity; likewise, we will explore opportunities for marijuana to improve its record. There's no reason marijuana must be environmentally unsustainable—consum-ers must simply become aware of the problem and demand action to correct it.

The first factor influencing hemp's agricultural sustainability is the intensity of exogenous inputs required to grow the crop. This can include pesticides, fertilizers, weed killers, and irrigated water. Like any crop, cannabis can be affected by stresses like animal and insect predation; plant pathogens like molds, viruses, and bacteria; or weeds that compete with the crop for soil nutrients for growth. Cannabis has a reputation for being a pest-free crop, for good reason. It has several natural features that help ward off predation, but it cannot ward off everything. It's better to think of cannabis as a pest-tolerant crop; there are around three hundred insect pests that attack cannabis crops, yet very few cause crop losses (McPartland 1996b, 52). This ability to tolerate pests is attributed to the insecticidal properties of its glandular trichomes, essential oils, and cannabinoids. Similarly, cannabis crops display natural resistance to diseases due to the natural biocidal properties of cannabinoids and terpenes. Nevertheless, some research suggests that upward of 11 percent of hemp fiber crops are lost to disease and drug crops may suffer even higher losses (McPartland 1996a, 19). The density of flowering tops of drug-type crops can be a suitable breeding ground for molds and viruses adapted to attack it, especially in humid environments.

Crops under pressure from pests or diseases require pesticides, which are generally toxic to both the targeted predator and to any other creature that encounters it. Since these pesticides do not discriminate between the harmful pest and the benign organism, they can dramatically reduce biodiversity, causing ripple effects in the environment. Hemp and marijuana differ in pesticide usage mostly due to pests' ability to damage flowers more than fiber or seed; since flowers fetch higher prices than fiber or seed, growers will take steps to assure the viability of their crop. This also applies to hemp varietals grown primarily for CBD content. Selective breeding for marijuana crops that display increased resistance to notable pests and diseases will increase the sustainability of these crops by reducing biocidal inputs.

Cannabis displays prolific growth characteristics, especially the "Sativa" biotypes favored by the marijuana trade. Fueling this growth often requires large inputs of fertilizers. When applied under optimal conditions fertilizers pose fewer environmental risks, but these ideal scenarios seldom occur. Rainstorms and flooding can wash out fertilizers into local water systems, causing damage to local vegetation and eutrophication in lakes and streams. As this happens at an increasingly large scale, aquatic systems will become unbalanced, resulting in excessive algal growth that starves any other organism of oxygen. This also creates ripple effects if these systems are interconnected—as one system chokes and dies it increases the likelihood that downstream systems will suffer the same fate. Hemp farmers commonly use manure as a fertilizer due to hemp's ability to rapidly take up nitrogen and its disposition for effectively utilizing animal wastes. Manure application can lead to runoff but, from an environmental perspective, is generally regarded as a safer fertilizer than petrochemical fertilizers. Marijuana poses greater risks, especially for urban greenhouses that discharge wastewater into municipal systems. Cities provide guidance on how to discharge effluent into municipal systems, so for the impact of marijuana growing to be effectively managed producers should comply with municipal ordinances.

Cannabis's prolific growth also means fiber crops do not require weed killers to manage other competing foliage, an additional benefit if seeking environmentally sustainable crops. Farmers plant hemp grown for fiber and seed very densely; the combination of dense planting and quick growth leaves little available nutrients or sunlight for competing weeds to grow. The canopy fiber hemp makes is so dense that few if any plants can photosynthesize and survive. In fact, some farmers in Canada use fiber hemp crops as a rotational crop to help naturally manage weeds on their farms. Instead of frequently spraying herbicides for weeds, the farmers simply plant hemp every few seasons to kill off any competing weeds. Marijuana and high-CBD hemp cultivars are not grown as densely, so they do not confer the same benefits as fiber hemp crops. Should cannabis breeders create crops that can be planted as densely as fiber hemp and can produce high cannabinoid contents in the flowers, this would be a huge future sustainability win.

While cannabis does require water in its early development, it is generally considered a drought-tolerant plant. This can vary between climates and weather patterns; however, its deep root structures break up compacted soil, allowing for better water retention, and the glandular trichomes help prevent wind desiccation later in the growing season. These useful characteristics allow for it to be grown in irrigated farmland, such as the eastern plains of Colorado or the Yakima Valley of Washington, without requiring extensive late-season watering.

Following the theme of water usage, cannabis does confer substantial water savings when compared to replacement crops. For example, according to a 2005 study, hemp requires a quarter of the amount of water than does cotton to produce a kilogram of usable clothing fiber and requires half the land to grow the same amount of finished material (Cherrett et al. 2005, 21). Replacing cotton fiber with hemp fiber would allow us to return more farmland back to natural wild habitat and save freshwater. Considering agriculture is responsible for 42 percent of freshwater usage in the US, this presents a significant opportunity (USDA 2021). Likewise, hemp products also present opportunities to replace nonrenewable products such as plastics and synthetic fibers. It is no surprise that fossil fuels require a significant amount of energy to be extracted, shipped, and refined into plastics, not to mention the direct environmental pollution caused through these activities. It requires significantly less energy to turn hempseed oil into plastic products. Hemp also presents the opportunity to create usable products and directly remove carbon dioxide from the environment. Products like hempcrete and, to a lesser extent, reinforcement fiber for traditional concrete can directly capture and store carbon dioxide into its structure. This presents significant upgrades from existing construction materials that traditionally have to be mined and processed, creating more environmental waste and carbon emissions.

When you combine the overall reduction of inputs used for growing hemp crops, combined with the added benefit of replacing a less sustainable alternative, there's already a compelling argument to consider hemp products. Hemp crops additionally allow us to not harvest from existing ecosystems, providing an additional environmental bonus. Take, for example, producing paper or particleboard from timber versus those same products produced from hemp. Research shows that one acre of hemp produces the same amount of fiber pulp as four acres of timber, while producing an extra crop of fiber or seed to boot (Small 2017, 411). This maintains the stand of trees, improving local biodiversity and sequestering atmospheric carbon dioxide. Extrapolated further, increased usage of hemp as an alternative to timber products can help us begin the process of reforesting unused land with more biodiverse woodland that will sequester even more carbon dioxide and, in time, become established, old-growth forest. This would be especially valuable in areas of the world where timber is in scant supply and deforestation is widespread. Hemp can help us balance needing managed forests grown for timber products and more established forest habitats for environmental benefits. Even if we supplemented part of our current timber usage, we could more sustainably manage our forests and not exacerbate the global trend of permanent deforestation.

Another exciting possibility comes from augmenting animal feed with hempseed, especially in aquaculture. Fish, like humans, need the essential omega-3 and omega-6 fatty acids contained in hempseed. Currently, we dredge the ocean for small baitfish to feed our farmed fish. This not only kills off local fish populations and disrupts food chains, but the trawling activities often also destroy reef habitats so fish cannot return. The use of hempseed to augment fish meal would go a long way toward keeping ocean populations intact while feeding humans.

I hope you agree that the potential global benefits from transitioning to greater use of hemp are worth the effort. For now, they merely exist as potential. Hemp is still a minor crop on a global scale and the supply chains needed to make it a major crop do not exist. We must all work to develop the infrastructure to achieve our goal of a more environmentally sustainable future. As I've stated earlier, this is one of the main reasons I wrote this book: to spur ideas about how to make hemp a major global crop so we can reap the environmental benefits. By doing so, brewers should look to use the lucrative parts of the plant so that economies of scale can kick in and make utilizing the other parts of the plant economically viable. For the most part, that will focus mainly on effectively harnessing the cannabinoids and terpenes present in female flowers of the cannabis plant. Thus, except for a brief look in the next chapter into the myriad uses of products derived from *C. sativa*, the remainder of this book will focus on how best to use cannabinoids and terpenes to develop compelling cannabis-infused beverages.

Before we move on, I would like to highlight a few key pieces of information generated by the USDA about the challenges hemp is facing in its bid to become a major crop. I would like for you to keep these challenges in the back of your mind when reading the rest of this book, so that you have a more thorough context of why this work is important and why we shouldn't be satisfied with just seeing a large influx of cannabis-infused foods and beverages. In its 2020 report, *Economic Viability of Industrial Hemp in the United States: A Review of State Pilot Programs*, the USDA laid out the successes and challenges from the 2014 Agriculture Act's hemp pilot program (Mark et al. 2020). After hemp's official legalization in 2018, the USDA report was issued to provide insight in how to bolster the crop so it could achieve long-term success. The report only focused on the issues pertaining to industrial hemp; however, I think it is equally important to think about marijuana's long-term role in our agricultural system and how it might become a more sustainable cannabis crop.

As of the spring of 2020, the vast majority of the hemp crop, about 95 percent of the planned acreage, was planted with the aim of producing cannabinoids, with the rest of the acreage dedicated to fiber and seed production (Hubbard 2020, 2). This tracks with the results of several economic surveys showing that cannabinoid production in hemp crops was the largest source of profit for farms. The current challenge for farmers is the highly fluctuating commodity CBD price, which has fallen from thousands of US dollars per kilogram of CBD extract to around $100 per kilogram (Mark et al. 2020, 32). The USDA report shows that there are significant barriers to stabilizing the price of CBD, some of which must be solved by the market and some by the government. The government, in the USDA's opinion, must clear up the regulatory environment, setting clear expectations regarding testing, licensing, and labeling standards for CBD products. The state-by-state variability of these requirements was hindering investment in consumer CBD products, causing major fluctuations in supply and demand. On the market side, new hemp growers sought to cash in on a short-term supply shortfall and flooded the market with CBD concentrate, forcing the commodity price to plummet. Prices will remain depressed until the oversupply can be worked through and a reasonable annual rate of supply can be established.

From a strictly agricultural perspective, there were many additional challenges farmers faced over the years of the pilot program that, if they remain unresolved, threaten to disrupt the nascent hemp industry. For starters, there are significant information gaps about the characteristics of different hemp cultivars and results of growing the same cultivar have varied from region to region, not least the levels of THC. For an industry that must remain under stringent regulatory guidelines around the production of THC, this level of variability can mean the difference between making a profit or a loss in the growing year. Further, there is no current transparent and reliable identification program for hemp seeds, nor is there any information regarding their origin. This makes some level of intuitive sense where several hemp cultivars were illegally grown prior to the passage of the relevant agriculture acts: for all intents and purposes, these cultivars simply appeared on the scene following legalization with little documentation about their origins or agronomic characteristics. In other cases, farmers purchasing feminized seeds at a significant cost premium found out later in the growing season that a significant portion of the seeds planted were male. This led to crop losses and legal action between seed growers and farmers; considering there was little documentation of seed origins or breeding practices, neither party could reliably prove the seeds were not female.

As of 2019, the Environmental Protection Agency had not approved any pesticides, fungicides, or herbicides for use on hemp crops. While these crops generally do not need large amounts of such chemicals, it is important to establish acceptable protocols in the event of infestation events that must be managed. If not, farmers risk crop failure and significant losses for their business. Eventually the risks will outweigh the rewards and farmers will be unwilling to grow hemp.

Additionally, there have been issues with farmers and law enforcement officials due to the challenge of determining the differences between hemp and marijuana based on visual inspection. In short, it is impossible to tell the difference between the two based on sight and smell alone, so farmers have had to be proactive when working with law enforcement to prove that their crop complies with the 2018 Agriculture Improvement Act. Additionally, farmers have also had problems with thieves stealing their crop thinking it was marijuana. I can attest I have seen many signs in the Colorado countryside communicating that they are growing hemp and to please not steal their crop. Crazy stuff.

Ultimately, the US hemp market needs time and stability to build up the infrastructure necessary to deliver a predictably profitable crop. Hemp must contend with competitive crops that may return at a higher rate, plus they must compete against hemp imports from countries that have built more hemp processing infrastructure and can undercut prices. Canada, the European Union, and China established legal hemp markets decades ago and have a competitive edge on the price and availability of many common hemp commodities.

Should these issues be resolved, farmers will have a stable and reproducible crop to grow and make reliable profits from. At this point, it is then incumbent upon the vested stakeholders (e.g., CBD processors, downstream product manufacturers, and, to some extent, local governments) to help these operations establish additional infrastructure to process secondary products from the main cannabinoid crop. For most, this will be employing decortication technologies to process fiber and hurds. These secondary products will dramatically increase the agricultural sustainability of cannabinoid-dominant hemp crops and will open additional opportunities of innovation and entrepreneurship in hemp.

For now, let us dive deeper into the subject of essential oils, cannabinoids, and terpenes so we can begin to build a sustainable cannabis economy.

4

USES OF THE CANNABIS PLANT

This chapter will explore the four component parts of the plant: the flowers and foliage, stalk, seeds, and roots. While each component of the cannabis plant is interesting from a brewer's perspective and, surprisingly, may feature many novel applications for brewing, for all intents and purposes this chapter will focus on non-beverage products derived from industrial hemp. Later chapters will focus specifically on cannabis in brewing itself.

There's nothing unique about industrial hemp, it's just that, currently, far more hemp cultivar biomass is utilized by various industries compared with marijuana cultivar biomass. Should legal barriers to growing marijuana come down, there is plenty of reason to think that we will be using more components of marijuana, just like we use the stems, seeds, roots, leaves, and flowers of hemp. Just about all of the cannabis plant contains useful components; the highest count of its uses comes from a 1938 article in *Popular Mechanics* claiming there have been over 25,000 identified uses of cannabis (Windsor 1938, 238)! In this chapter, we will look at each of the main components of the cannabis plant in turn and talk about some applications that may be of interest to brewers. This review is by no means exhaustive.

We will start with a brief discussion of essential oils, which are the complex and varied mixtures of volatile organic compounds produced by all plants. Some plants produce more essential oils than others and are therefore of commercial interest as essential oil crops. *Cannabis sativa* is one such plant, producing essential oils in its glandular trichomes (we looked at the role of trichomes in chapter 3). Hops—species of the genus *Humulus*—are also plants whose essential oils are of great interest to humans, namely, brewers. Similar to cannabis, hops possess secretory, or glandular, trichomes. In fact, two components in each plant, cannabinoids and alpha and beta acids, are analogous in that they share a common biochemical root: they are both prenylated terpenophenolics. Researchers have demonstrated that cannabinoids and hop bittering acids share common biosynthetic pathways, with implications for estimating when and how the two plants split apart in their evolutionary history (Page and Nagel 2006, 187).

The two common components we will briefly review here are the essential oils and the cannabinoids. These compounds, produced in the multicellular glandular trichomes, perform various roles, including (for the purposes of the plant):

- Antibacterial effects
- Antifungal effects
- Antiviral effects
- Antiparasitic effects
- Cytotoxicity in herbivorous insects and animals
- Attracting pollinators
- Attracting beneficial predatory insects
- Protection from ultraviolet light

ESSENTIAL OILS

There are many kinds of oils we will discuss in this book. The three main oils are: seed-derived oils (from pressed seeds), cannabinoid-rich oils (from glandular trichomes), and essential oils (also from glandular trichomes). Cannabinoids and essential oils are secondary metabolites derived from various products biosynthesized by the methylerythritol phosphate (MEP) pathway (sometimes known as the deoxyxylulose phosphate, or DOXP, pathway), mevalonate (MVA) pathway, and fatty acid metabolism. The MEP and MVA pathways are common biosynthetic pathways for producing compounds for cell membrane maintenance, hormone synthesis, and binding intracellular proteins. While cannabinoids and essential oils are not fundamental to the growth, development, and reproduction of *Cannabis*, these secondary metabolites play an important role as protective agents from predation and disease.

Essential oils are produced by many plants; they are the volatile aromatic oils that give plants their signature scent. Plants like hops and mint have secretory glandular trichomes in the same way that cannabis plants do. We can look to industries that use those essential oils to inform our understanding of cannabis essential oil.

In order to accurately study essential oils, one must isolate them from the other constituents found in the glandular trichomes. Essential oil from a plant is a heterogenous and complex mixture of different compounds, consisting mostly of organic hydrocarbons and their oxygenated derivatives. These include multiple classes of terpenes[1] and terpenoids (terpenes with additional chemical units that include oxygen) and oxygen-containing compounds such as alcohols, esters, ethers, aldehydes, ketones, lactones, phenols, and phenol ethers (Ross and ElSohly 1996, 49). The volatiles comprising essential oils are either direct products of plant metabolism or they evolve from oxidative reactions of volatile and non-volatile precursors (Rettberg, Biendl, and Garbe 2018, 3). Studying the composition of these oils, especially in the pursuit of reproducibly replicating these aromas in plants year after year, is maddeningly complex. Factors such as plant maturity, genetics, drying, packaging, storage conditions, and terroir (i.e., the confluence of geography, climate, and soil) all influence the final composition of a crop's essential oil profile. The odor of the freshly living plant is different from dried plant material (Ross and ElSohly 1996, 51), a phenomenon to which brewers who have walked into a hop-yard and then rubbed dried hops can attest.

While not directly involved in growth and reproduction, essential oils and cannabinoids perform additional protective functions for the plant. Through multiple studies, researchers found that many essential oil constituents and cannabinoids, particularly intermediary acidic cannabinoids, are toxic to many organisms (Radwan et al. 2008, 2631). Essential oils generally have varying levels of antimicrobial, antifungal, antiviral, and antiparasitic properties; most researchers agree that certain biotypes of

---

1 Terpenes are made up of five-carbon units called isoprenes (C_5H_8) that are joined together to form terpenes of varying lengths ($(C_5H_8)_n$), where *n* is 2 or more. The smallest terpenes are monoterpenes, formed by the joining of two isoprene units to make a ten-carbon (C_{10}) molecule. Larger terpenes include sesquiterpenes (C_{15}), diterpenes (C_{20}), triterpenes (C_{30}), and so on.

cannabis more than likely evolved as they created certain terpenes to ward off specific predators in their environment. In this light, it makes sense that the cannabis plant's glandular trichomes are concentrated around its reproductive tissues.

Additionally, researchers and growers have demonstrated that some terpenes attract predatory insects that help control pests (McPartland and Sheikh 2018, 1296). Many organically farmed or low-input cannabis farms stress the benefits of these characteristics as an essential part of their integrated pest management programs.

Cannabis essential oil is easily obtained by steam distillation, vaporization, or extraction with a non-polar solvent. Yield quantities depend on the *Cannabis sativa* "strain" or "land race"—cannabinoid-rich strains yield more essential oils, fiber-rich strains yield less—and whether the female was pollinated. Male plants do possess small numbers of glandular trichomes, though it is generally not economical to harvest them for that purpose. Additionally, harvest time, storage, and drying conditions all factor into the yield and composition of the essential oils of cannabis. Traditionally, harvested cannabis is hung and dried with fans that create air circulation. We will discuss quality considerations for hemp as a brewing ingredient in chapter 10.

APPLICATIONS OF ESSENTIAL OILS

Insecticides

Researchers have studied cannabis essential oil's ability to function as a natural insecticide, which is in addition to its better-known properties as an antifungal and antimicrobial agent (Benelli et al. 2018, 309). In this study, the researchers prepared a steam-distilled essential oil from fresh cannabis and then applied it as a spray to a variety of insects. The oil was tested against mosquitos, peach aphids, houseflies, and the tobacco cutworm (a moth larva) to test toxicity; it was also tested on ladybugs and earthworms to show it is not toxic to beneficial invertebrates. The test insects represent many common pest types that are detrimental to society as disease vectors or crop destroyers. The study showed varying degrees of invertebrate toxicity but did conclude that cannabis essential oil could be effective as an insecticide in organic agriculture or integrated pest management programs, especially for the management of houseflies and aphids. Should the economics of harvesting cannabis for its essential oils improve, this could be a very exciting application as a lower-environmental-impact insecticide.

Organoleptics

Terpenes garner lots of attention in the marijuana community, for good reason. Researchers, both professional and amateur, purport that terpenes drive the "entourage effect." The entourage effect theory states that all the individual compounds work synergistically with one another when consumed and give the most efficacious result, be it for medicinal or recreational effect. I describe it as a theory because it has yet to be proven conclusively in clinical research. Further discussion on the theorized role of terpenes in the entourage effect can be found in chapter 8.

Terpenes dominate most of the essential oil profile, with monoterpenes making up 48–92 percent of the total oil composition by weight and sesquiterpenes making up an additional 5–49 percent, depending on the biotype. Common terpenes in cannabis include myrcene, β-caryophyllene, α-pinene, *trans*-ocimene, and α-terpinolene (Mediavilla and Steinemann 1997, 82). No one terpene drives the characteristic cannabis aroma, but many researchers study the aroma composition for purposes ranging from law enforcement detection to organoleptic sensory analysis.

Drug dogs are trained to specifically detect β-caryophyllene when doing searches for marijuana. β-Caryophyllene is one of the few terpenes consistently found across all cannabis and hop cultivars.

We as brewers know quite a lot about terpenes through our use and study of hops in the brewing process. Using a similar approach to our study of hop chemistry, we can analyze several available strains of hemp flower to better understand the composition of their essential oils. We will talk about how to manipulate these compounds in the brewing process later in chapter 10.

CANNABINOIDS

Cannabinoids are the other primary compounds synthesized in the glandular trichomes of cannabis. We primarily associate cannabinoids with the most notorious one identified, delta-9 tetrahydrocannabinol (THC), which gets us high. In fact, there are over 100 identified cannabinoids produced by cannabis, with more being discovered every year. Most cannabinoids do not get humans high but do produce varying levels of physiological effect when ingested. While cannabinoids are not wholly unique to the *Cannabis* genus, they are predominantly found in *Cannabis sativa*. Certain species of *Rhododendron* and *Helichrysum* (a type of daisy) also produce cannabinoids and several other plants produce compounds that mimic the effects of certain cannabinoids (Appendino et al. 2008, 1427).

Cannabinoid concentrations in the cannabis plant increase from the initial formation of flowers, reaching peak concentrations at maturity, the timing of which varies depending on environmental conditions. The production of cannabinoids is correlated with the production of terpenes; terpenes serve as a biosynthetic intermediary for cannabinoids (Booth, Page, and Bohlmann 2017, 4). Relatively speaking, as the concentration of cannabinoids increases, the concentration of certain terpenes will decrease.

Similar to terpenes, researchers posit that cannabinoids play important plant-protective functions. The evolutionary origins of cannabinoids are still subject to debate, though Ernest Small puts it succinctly:

> All of the preceding does not provide a definitive answer to the question of why the cannabinoids have evolved, an issue that remains open to speculation (indeed, why other species in the Cannabaceae have secretory epidermal cells is equally unclear). Most secondary compounds are likely (a) metabolic waste products, (b) generalized antibiotics (acting against all harmful classes of organisms), or (c) evolutionary holdovers from ancestors in which the chemicals were adaptive. (Small 2017, 219)

For at least the last 100,000 years, *Cannabis sativa* adapted to produce cannabinoids in abundance because of human selection and domestication.

Cannabinoid production can be manipulated by environmental factors, especially factors related to plant stress. Changes in light, warmth, water availability, soil nutrition, and carbon dioxide concentration will all alter cannabinoid production, with stresses on the plant increasing the overall cannabinoid content. These stress responses reinforce the theory that cannabinoids play a role in protecting the cannabis plant's reproductive system from environmental stresses like drought, desiccation, ultraviolet light damage, and disease. Research on human disease further reinforces these properties; cannabinoids like cannabigerol have been shown to be effective against bacterial and fungal infections such as MRSA and a variety of molds (Appendino et al. 2008, 1430).

Cannabinoid production is also influenced by genetic factors. The only common cannabinoid produced across all *Cannabis* plants is cannabigerol (CBG), which is considered the "mother" cannabinoid. From there, varying quantities of either tetrahydrocannabinol (THC), cannabidiol (CBD), or cannabichromene (CBC) are produced. Typically, marijuana varieties produce an abundance of THC with little to no CBD or CBC, and hemp varieties produce abundant CBD with little to no THC or CBC. This is an oversimplified way of looking at it, but it's a suitable stand-in until we discuss the topic in more detail in chapter 5.

One curious property about cannabinoids is that, although we associate them with their psychoactive and pharmacological effects, cannabis plants do not naturally produce cannabinoids in an "active" form, that is, one that will produce an effect in humans. Naturally occurring cannabinoids exist as acidic precursors and, if ingested raw, will not get you high. Cannabinoids must be heated to become psychoactive.

PRODUCTS DERIVED FROM PARTS OF THE CANNABIS PLANT

Inflorescences and Trichomes

It is easier to think about cannabis plants, especially the inflorescences, as trichome factories. There's nothing special about the vegetative plant matter from a human use perspective, it is simply the vessel that carries the good stuff. Cannabis shares a lot in common with its cousin, the hop. We don't particularly care about the vegetative matter in hop plants, we simply want the trichomes on the hop cones that deliver the bitterness and aroma we are looking for in our beer. In fact, we often design complicated processes to remove the vegetation from hops to concentrate the oils found in hop trichomes.

Obviously, the most common application for cannabis is as a recreational and medicinal drug. Humans have been getting high from *Cannabis sativa* in various forms since the dawn of civilization. Recently, we have figured out new ways to extract and infuse its constituents into novel products, including vaporizers, waxes, shatters, beverages, foods, candies, pills, and powders. This book will focus mostly on the beverage applications of cannabinoids; a review of these technologies can be found in chapters 6 and 7. There are many excellent additional resources to learn more about the psychoactive and medical constituents of cannabis trichomes, including the companion book to this volume, *Brewing with Cannabis: Using THC and CBD in Beer* by Keith Villa.

Seeds

The next useful component of the cannabis plant are the seeds. Seeds grow out of the perigonal bract of the flowers after a female plant successfully reproduces. The density of seed production is correlated to the density of the flowers; certain *Cannabis* cultivars will produce more seeds than others. Typically, farmers grow hemp varietals to produce seed, so we will refer to the seed from here on out as hempseed. The predominant reason for growing exclusively hempseed is a legal one: hemp cultivars enjoy greater legal status than marijuana strains. As discussed in chapter 3, the perigonal bracts contain the highest concentration of cannabinoids, so seeds can become contaminated with THC during harvesting and processing should the THC concentration be high on the plant.

Hempseed is an oilseed, sharing similarities with flax, canola, sunflower, and soybeans. In contrast to those crops, hempseed is remarkably nutritious and contains many constituents suited to a healthy, balanced diet (Callaway 2004, 65). The seed (technically an **achene**) of a cannabis plant is oblong in shape, roughly two to four millimeters wide, and about three to six millimeters long. The seeds can range in color from a light-yellow hue to a deep forest green. It is enveloped by a thin, brittle shell that contains little nutritional value and is typically removed before processing.

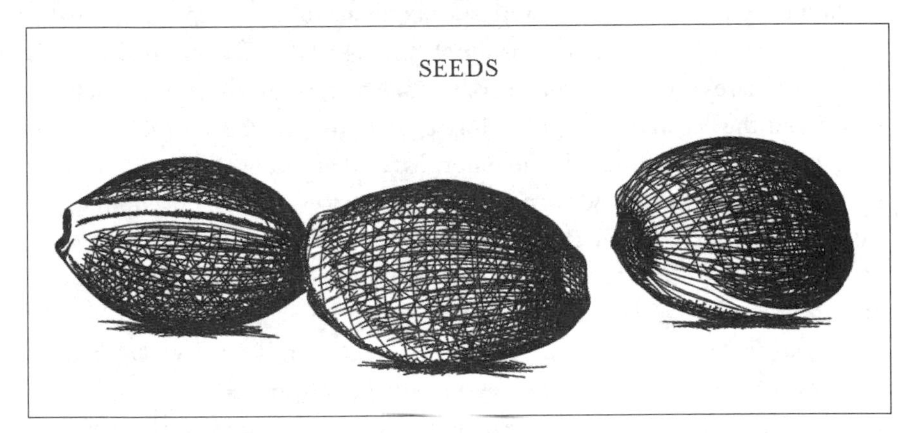

Figure 4.1. *Cannabis* achenes ("seeds"). © *Hidden Powder Design*

Hempseed's nutritional value has been recognized by various cultures for several millennia. Due to some of the processing challenges with hempseed and its propensity to turn rancid under poor storage conditions, many cultures moved away from eating hempseed as they developed more shelf-stable crops. It has always remained a nutritional backstop for many farming cultures and was prized as a food during times of famine. Hempseed contains roughly equal quantities of carbohydrate, protein, and fat. Additionally, it also contributes dietary fiber and roughage, as well as antioxidants and minerals, including calcium, iron, magnesium, phosphorus, potassium, sulfur, and zinc (Pate 1999, 247).

Hempseed is best known for its nutritious fatty acid content. Proponents of hempseed's nutritional benefits often describe fresh hempseed oil as "nature's most perfectly balanced oil" containing an optimal balance of omega-6 and omega-3 fatty acids (Small 2017, 150). The oil contains primarily unsaturated fatty acids, including oleic acid (16%), omega-6 linoleic acid (55%), and omega-3 linolenic acid (25%), with lower concentrations of saturated fatty acids like palmitic acid (9%) and stearic acid (3%). In humans, healthy diets containing balanced omega-6 and omega-3 fatty acids generally result in lower cholesterol levels, decreased risks for cardiovascular disease, and decreased levels of inflammation. Humans do not naturally create omega-6 and omega-3 fatty acids and must get these essential fatty acids through their diet. Many people take supplements derived from fish oil to help balance their fatty acid intake; hempseed is a viable replacement for those supplements.

Hempseeds are most often eaten whole, sometimes with the shell removed. While the shell does not contain much nutritional value, it protects the seed from oxidation so it is often left intact if shelf life is a concern. When hempseeds are dehulled—the shell removed—the remaining seeds are called hemp hearts, or hemp nuts. They have a pleasant nutty flavor, comparable to a mix of almonds and sesame seeds, and are quite mealy in texture, which makes them suitable for adding to trail mix, granola, oatmeal, and baked goods. We will explore how to best utilize hempseed as a cereal adjunct in beer in chapter 9.

In addition to being eaten whole, hempseeds are often cold-pressed into cooking oil. The pressed oil retains much of the flavor of hempseed and lends a light herbaceous note to salad dressings, mayonnaise, and even baked goods. When heated, hempseed oil has a low smoke point, roughly 329°F (165°C), which makes it unsuitable for frying applications. The residual seed meal from the pressing process, called hempseed cake, can be further processed into gluten-free flour for baking or extracted into a plant-derived protein powder. If the hempseed cake is not used for human food, it is often sold as a prized animal feed. Ornithologists recognized for years that hempseed is an excellent bird food and some poultry farmers feed their chickens with hempseed to enrich the eggs with omega-3 fatty acids. These applications tend to be sporadic for now, since the hempseed crop has not reached a large enough size to be economically commoditized.

Historical legal prohibitions on cannabis suppressed the growth of hemp, including hempseed. Since hempseeds can contain detectable quantities of THC if not cleaned, the US government took a deeply skeptical view of hempseed's nutritional benefits for decades. If you are old enough to remember, pro-athletes and military personnel used to blame eating hempseed for failing drug tests. There's an element of truth to this: cannabis seeds can contain as much as 200 micrograms of THC per gram of seed, depending on the cultivar (Ross et al. 2000, 717). Most hempseed farmers are required to clean their seeds of cannabinoids before they can sell them; we will discuss this issue and further quality concerns in chapter 11.

In addition to being used in food, hempseed oil can also be used for making soaps, biodiesel, paints, varnishes, cosmetics, and lotions. Throughout history, hempseed oil was also burned as lamp oil and was a cheaper alternative to whale oil. Hempseed oil burns very dirtily, however, so it was used when other oils were too expensive or scarce.

Fiber

As we move further down the plant, the next useful component of cannabis are the fibers. You may associate hemp fiber with rope, or the necklaces that hippies wear, but there is considerable diversity among the qualities of fiber produced by the cannabis plant. There are two basic classes of fiber that can be obtained from the stems of cannabis: the phloem (bast) fibers and xylem fibers (woody fibers, or **hurds**).

Both fiber types are associated with the vascular fluid transportation system of the plant and are the primary components of the stalk and branches of the plant. The woody fiber tissue constitutes the core of the stalk and transports water and solutes from the roots to the other parts of the plant. The bast fibers transport photosynthetic metabolites from the foliage to other parts of the plant for normal growth and maintenance.

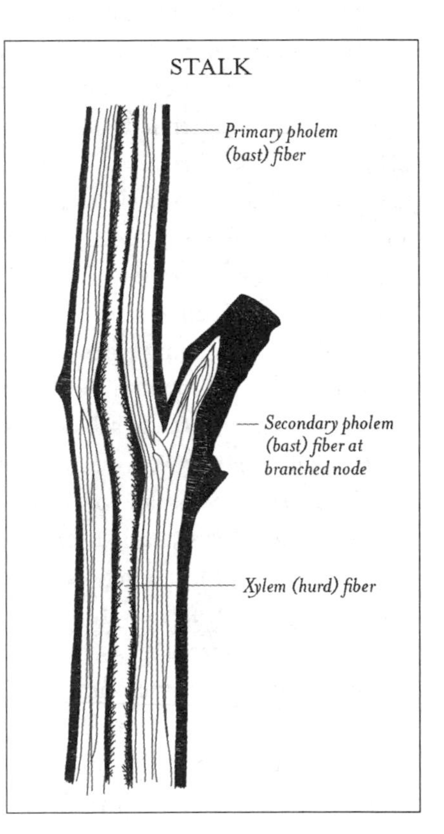

Figure 4.2. *Cannabis* stalk, showing its different types of fiber. © *Hidden Powder Design*

The reason why cannabis fibers are so strong relates to its evolution: it grows very tall in open, treeless habitats and reproduces via wind pollination. Thus, it needs to be able to withstand heavy winds without tipping over. The woody fibers, or hurds, provide the main vertical strength of the plant. The bast fibers provide the stem with elasticity, allowing it to bend and buckle without breaking.

There are both primary and secondary bast fibers, each providing their own unique benefits to the plant. The primary fibers are longer and stronger than secondary fibers and provide the most structural support for the plant. The primary fibers are the most prized fibers for human applications, which we will get to in a moment. The secondary bast fibers lend strength to the primary bast fibers and are found in abundance around reinforcement points, such as the base of the plant and around the branch nodes. They tend to be thinner and shorter than the primary fibers. The bast fibers have interesting physical properties that allow for them to be strong yet comparatively light. They contain four organic polymers that all work synergistically together: cellulose, lignin, hemicellulose, and pectin. Cellulose provides the main structure to the fiber and has high tensile strength; lignin binds to cellulose, providing additional structure. Hemicellulose binds both cellulose and lignin, providing even more reinforcement. The bound cellulose, lignin, and hemicellulose form a tube that extends for most of the fiber. To form a complete bast fiber, pectin binds the individual tubes of cellulose, lignin, and hemicellulose. Pectin has a gummy consistency that provides much of the flexibility in the plant fiber.

Based on tonnage, hemp currently constitutes only about 0.3 percent of the world's natural fiber production, excluding wood fibers (Small 2017, 91). China produces the most processed hemp fiber in the world, followed by the European Union and Russia. As we discussed before, the market demand for natural fibers is quite low, due to the availability, cost, and performance of synthetic alternatives. As we continue to better understand how to use synthetic and natural fiber products, opportunities will arise for both to compete, with their economic utility weighed against their environmental impact. With the increasing weighting of environmental impacts on bottom lines and with the increasing acreage of hemp, natural hemp fiber products will increasingly be able to compete. Along with understanding the natural properties of hemp fibers, it is equally important to understand how we process the stalks to extract fiber from the bast and hurds. There are now several modern products that can be created from hemp that extend the uses of the cannabis plant beyond just textiles, cordage, and specialist pulp for paper.

Humans figured out how to efficiently extract fibers from hemp about as long as they have been using it. The easiest way is to simply peel the outer fibers off the base of the stalk, either fresh or after it has dried for a little while after cutting the plant down. Early humans more than likely employed these fibers to stitch together animal hides or other crude forms of clothing. They quickly realized, however, that they can vastly improve the quality of the long fibers of hemp through a process called **retting**. Retting, a word meaning "to rot", is a method of bacterially and enzymatically breaking down certain plant fibers. The traditional method of retting involves soaking hemp fibers in water and allowing natural bacteria and fungi to break down the pectic matrix, freeing the long fibers for further processing. In certain climates, a farmer could "dew ret" hemp stalks by simply piling them and allowing rainfall to sufficiently saturate them.

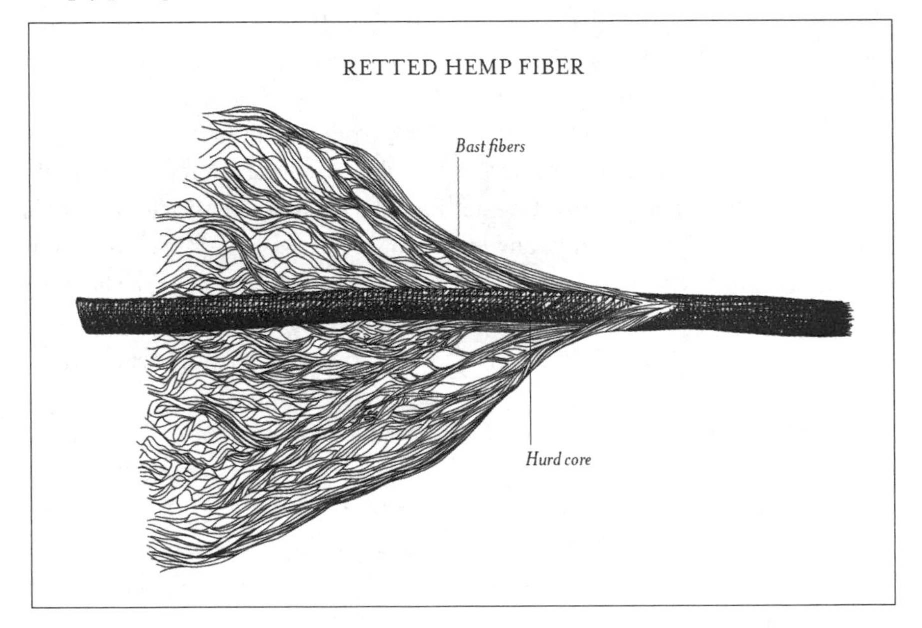

Figure 4.3. Retted *Cannabis* stalk showing the bast fibers and hurds. © *Hidden Powder Design*

Throughout most of history, and currently in certain parts of the world with lax environmental regulations, farmers would submerge their hemp crop in a nearby ditch, creek, or pond for retting. "Water retting" in this fashion leads to high levels of water pollution and the bacterial fermentation leads to such a low oxygen saturation in the water that it kills any wildlife living there. China, unfortunately, still practices this type of retting as a form of convenience and cost control and as such controls most of the hemp fiber and textile trade globally (Zhang et al. 2008, 1199).

Modern forms of water retting involve large tanks that circulate water and have temperature control to optimize bacterial growth and the speed and consistency of retting. These plants also have sophisticated wastewater treatment facilities to comply with water quality regulations, making them expensive to build and maintain. Other retting techniques, such as chemical processing, thermal retting, steam explosion (Garcia-Jaldon, Dupeyre, and Vignon 1998, 251), or enzymatic retting (Pakarinen et al. 2012, 275), have been developed to lessen capital requirements but have not seen broad-based adoption.

Once the stalks complete the retting process, they must undergo the next step in processing: **decortication**. Decortication is the process of removing and sorting the fiber at the base of the stem from the inner woody core. Decortication technology has not changed much in the last 150 years, although the quality of both modern decorticators and their output products has improved dramatically. The end of the decortication process results in three different materials, or four with modern decorticators: long fibers, short fibers, hurds, and high-cannabinoid waste (from modern decorticators).

The first step in the decortication process is to break the stalk to loosen the outer long fibers from the base. Rollers, twisters, and other mechanical tools facilitate the loosening and begin to break apart the core into smaller woody pieces. From there, the long fibers need to be further separated from one another using a process called "scutching" (Small 2017, 108). The scutching process, sometimes done by hand but more often mechanically, involves gently beating the fibers against a series of cleaning knives to remove any impurities from the fibers. Once the decorticator removes the impurities from the fibers, the fibers can be sorted by length and quality for further processing. With the rise in commercial demand for cannabinoids—mainly CBD—decortication technology has added a further step to reclaim the waste "dust" that contains small quantities of glandular trichomes. While this waste stream is not as efficient as extracting cannabinoids from the leaves and inflorescences, it is can capture a meaningful quantity of cannabinoids for extraction as a bonus revenue stream (Christensen 2019, 2). Modern engineering has developed mechanical decortication solutions that provide superior results, increasing both yield and quality. Fresh green stalk or dried bales (or even retted material) can be processed without the time, energy, or waste usually involved with conventional methods.

Construction Materials from Hemp Fiber

Hemp-reinforced concrete. One of the most exciting products derived from hemp fiber comes from the construction sector. Durable hemp bast fibers can reinforce concrete to prevent common forms of failure: cracking, warping, and breakage. Replacing defective concrete makes up a significant amount of its environmental impact. Adding reinforcing fibers to a concrete mix can increase the usable life of traditional poured concrete. The use of hemp in concrete structures dates to the Roman Empire; in fact, there are still hemp-reinforced concrete bridges in Europe today. If you've poured a brewery floor, you've likely used fiberglass or some other binding product to make sure your floors set well and resist cracking. The next time you need to pour a new floor, consider hemp.

Hempcrete. Hemp fiber can not only reinforce existing materials that you're familiar with, but can also make a product that, in my opinion, should be in every brewery: hempcrete. Hempcrete is a fiber-reinforced material made from a combination of hemp hurds, powdered lime, and water. It's easiest to think of hempcrete as a combination of drywall and insulation, but it has the appearance of concrete. Despite the name, hempcrete cannot be used as a replacement for concrete in structural or heavy load-bearing applications, it is primarily used for the infill of a building. Despite these downsides, hempcrete has two significant advantages as a building material that should be of interest to brewers: it's highly insulative and it absorbs more carbon dioxide during its manufacture than it emits.

Hempcrete is highly insulative to temperature and sound and regulates temperature and humidity in any indoor environment. This has huge potential for large buildings requiring good insulative properties, as we will see below. Proper insulation is essential for energy efficient buildings, but hempcrete's best aspect is that it's carbon negative. Hemp absorbs carbon dioxide during its growth cycle and stores it in its biomass when harvested. When the hemp biomass is mixed with the water and lime binder, the lime calcifies the embodied carbon dioxide in the hemp biomass into calcium carbonate. The amount of carbon dioxide sequestered depends on the ratio of materials within the hempcrete, but repeated studies show that hempcrete could be a significant carbon sink for the future.

In a UK study, Ip and Miller (2012) found that for every square meter (~11 sq. ft.) of hemp-lime wall at a thickness of 0.3 m (~12 inches) without any wall finishes can sequester 82.7 kilograms of carbon dioxide. They also studied the amount of carbon emissions generated by constructing the wall: everything from growing and harvesting to processing to shipping and, ultimately, making the hempcrete. They found that the entire process in the hempcrete manufacturing and supply chain generated 46.4 kilograms of carbon dioxide, meaning that the net carbon dioxide storage was 36.1 kilograms per square meter (Ip and Miller 2012, 8).

A similar study conducted in Ireland (Walker and Pavía 2014, 275) showed that one square meter of hemp-lime wall 260 mm (~10 inches) thick required 370–394 megajoules of energy for production, compared to 560 megajoules for an equivalent section of Portland-cement concrete wall. What's more, the hemp-lime wall was calculated to *sequester* 14–35 kilograms of carbon dioxide over its 100-year life span, whereas the Portland-cement concrete wall *released* 52.3 kilograms of carbon dioxide. The Walker and Pavia study highlights that not only is hempcrete carbon negative, but if used in place of a net emitter of carbon dioxide like traditional concrete the savings in carbon dioxide emissions become even more impressive. One study found that the construction sector globally was responsible for 5.7 billion tons of carbon dioxide emissions in 2009, which was 23 percent of all carbon dioxide emitted from human activity that year (Huang et al. 2017, 1906). Ninety-four percent of those sectoral emissions came from indirect sources, mostly the materials themselves. Traditional concrete alone was the single largest contributor to these indirect emissions, contributing 14 percent of the indirect carbon dioxide emissions from the global construction sector.

Additionally, the materials used for the construction of a building make up only part of the energy intensity of the space. The energy required to operate the space makes up the balance and is ongoing for the life of the building. Selecting low-energy materials helps lower the embodied energy, in some instances by more than 50 percent, but those materials also need to make efficient use of the ongoing energy inputs used to heat, cool, and power the building (Venkatarama Reddy and Jagadish 2003, 129). Hempcrete is highly insulative and is a good resistor of thermal conductivity (depending on the binder material and thickness of the wall); because of the small pockets of trapped air in hempcrete, very little heat is lost through the wall (Walker and Pavía 2014, 274).

You may be asking yourself at this point, why isn't hempcrete easily available and used everywhere? The easy answer is that the infrastructure for supplying enough raw material is, at best, underdeveloped and even in places where you can find enough raw material there are few skilled tradespeople who can correctly install it. In other words, it's expensive. Another barrier in the US is that the American Society for Testing and Materials (ASTM) has yet to certify hempcrete as a measured and approved building material. At the time of writing this, the ASTM is actively developing standards for the fire rating and thermal resistance (insulation) score of hempcrete so it can provide guidance for municipal zoning and permitting agencies. Hopefully, industries that need to insulate large buildings—the brewing industry, for one—will consider this material in the future as cost and regulatory barriers are reduced.

Unlike in the US, the European Union has already cleared these hurdles and embraced hempcrete as a viable construction material. Adnams, the brewery in the UK, built a distribution center entirely out of hempcrete in

2006. A case study conducted by the project's construction firm estimated that the 4,500 square meter distribution center directly sequestered 150 tonnes of atmospheric carbon dioxide and prevented over 600 tonnes of carbon dioxide from being emitted should they have gone with traditional building materials (LimeTechnology 2006). Additionally, the distribution center is so thermally efficient it does not require any mechanical cooling. Hempcrete presents opportunities for brewers building large, thermally efficient buildings to substantially reduce not only their environmental impact but also their continuing operational costs. This opportunity should be treated seriously by brewers.

Fiberboard. Another interesting product that uses hemp hurds is particleboard and fiberboard. Manufacturers design fiberboards to fit a variety of specifications, but they are most widely used as wallboards or as a base for furniture. Many furniture manufacturers flatpack and ship their products to be assembled at home by the consumer, an experience anyone who has shopped at IKEA can attest to. This type of furniture requires centralized manufacturing coupled with extensive shipping and distribution, which can be economically cheap yet environmentally expensive.

Traditionally, fiberboard contained waste wood particles from lumber manufacturing, but alternative feedstocks have been introduced in recent years. China is the current leader in developing alternate inputs for fiberboard manufacturing due to the country's residential and industrial manufacturing boom and its inability to find sufficient sources of lumber to meet demand.

Several studies showed that hemp hurds and bast fiber are suitable raw materials for the manufacturing of fiberboard (Schöpper, Kharazipour, and Bohn 2009, 368). Despite being lighter than poplar or beech wood, hurds have a higher bond rating and superior bending strength when compared with a standard wood fiberboard (Nikvash et al. 2010, 325). With the higher bond rating, hemp fiberboard was not only suitable as a replacement for traditional wood fiberboard but in fact scored higher in performance and durability. The lower density and weight means that manufacturers can reap additional benefits by reducing energy and transportation costs. You don't have to manufacture fiberboard with only hemp-based feedstock either—a study showed that using hemp in conjunction with traditional wood conferred many of the same benefits of reducing overall weight and improving strength while still maintaining or surpassing existing regulatory standards (Li et al. 2014, 187).

Currently, the largest challenge in the particleboard industry is the cost and quality of wood to manufacture particleboard. Consolidation of lumbermills has concentrated the supply of lumberyard waste into the hands of a few companies and this often requires longer transportation times to individual particleboard manufacturers. Additionally, the rise of biomass production for use in the energy sector, often in the form of wood pellets, creates competition with traditional manufacturers and drives up the price of raw materials. This, of course, is a retrospective view of the problem and does not consider the increased importance of maintaining and growing forest cover in the coming years to combat atmospheric climate change, sequester carbon dioxide, and protect biodiversity.

Hemp, with its ability to grow quickly, could become a potential solution to this burgeoning problem. Currently, it is not economical to manufacture hemp-based particleboard, but as increased acreage and processing infrastructure come online this could be a promising way to produce a superior product while reaping bonus environmental benefits.

Other Products from Hemp Fiber

Aside from construction materials, there are literally thousands of additional hemp-fiber products that should be of interest to brewers. I won't belabor the point, but I think it is useful to call back to the hempcrete study that compares its environmental impact compared to its traditional competitor. I think this is the most salient comparison we can make when thinking about more environmentally beneficial replacement products for breweries. Does your current cardboard come from cutting virgin forest? Maybe consider recycled paperboard or hemp-based paperboard. Are your festival cups recyclable? If they are, do you know whether they are actually recycled? Hemp bioplastics are not only recyclable, but biodegradable. Of course, we consider the costs along with the environmental impacts of our materials, and hemp-based products will undoubtedly be more expensive in the near term. If you can afford the extra expense, you will help drive down the costs for everyone later. The quicker we can adopt these materials, the quicker we can achieve a more sustainable world.

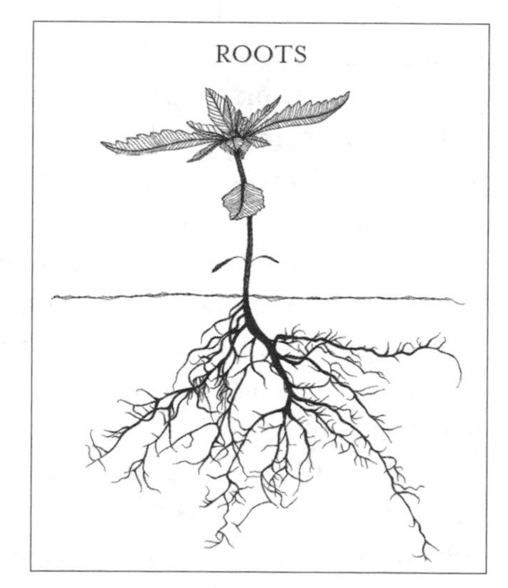

Figure 4.4. Profile of *Cannabis* plant root system. © *Hidden Powder Design*

Roots

As we reach the bottom of the cannabis plant, we get to an important, unsung hero: the roots. Cannabis evolved adjacent to river systems, so its natural preference is for well-drained, sandy-loam soils that have adequate nutrition. Cannabis owes its tolerance of a variety of soil types and its environmental adaptability to its root system. The moniker "weed" wasn't given to cannabis by accident; it will grow and thrive in most environments, even crowding out existing foliage. Cannabis's dense, quick-growing root structure allows it to gain a competitive environmental advantage. There are crucial differences in the growing characteristics of marijuana and hemp; hemp tends to be a much more forgiving and robust crop, whereas marijuana can be more finicky. The ideal soil type for hemp should have no more than 40 percent clay or 60 percent silt, as well as no less than 35 percent sand (Haney and Kutscheid 1975, 6). Sand will contribute to soil aeration and drainage.

There are quite a few anecdotes about hemp root systems that are mostly true but should not be applied in every situation. Much of the wild hemp population in the US exists in irrigation ditches, where it is either intentionally left to thrive or allowed to take over because of its ability to control soil erosion. Fiber hemp cultivars grow extremely tall and need deep, complex root systems to anchor them.

Root systems of hemp consist of a central taproot and multiple lateral branching root groups. In optimal soil conditions, the taproot can reach as deep as seven feet (2 m). This benefits soil health in a variety of ways, including breaking up deep soils to allow further drainage and resist compaction, transport of nutrients and minerals from deep soils, transport of water lowering the need for irrigation, and reducing topsoil erosion. With these benefits, hemp is a useful rotational crop that can improve the health of the soil, benefitting the farm for multiple seasons.

These benefits are not always as broadly recognized as many hemp proponents care to admit. Many hemp crops grown for CBD content show similar growing characteristics that marijuana plants often show in outdoor farms. These plants tend to be smaller, thus, they do not form as deep of a taproot as fiber hemp cultivars. This tends to leave shallow root systems that do not control soil erosion anywhere near as well as fiber hemp and do not yield the same benefits for soil health. Additionally, in areas with compacted or poor-quality soils, not only will cannabis not break up soils. Paul Clift, who operates a hemp farm in northern Colorado, reports that compacted soil can cause the taproot to wrap around itself, strangling and killing the plant (Paul Clift, pers. comm.).

Well-established root systems help soil in two primary ways: the control of soil erosion and decompaction. Cannabis farming requires the soil to be tilled each year, raising the risk of soil erosion. Wind and rain can sweep off the upper layers of topsoil, decreasing the overall fertility and structure of the soil. Thankfully, cannabis establishes root structures quickly, minimizing the period in which the weather can erode the soil. Cannabis roots are densest in the top five inches of soil, providing the most structural protection for topsoil (Amaducci et al. 2008, 227). Once established, the dense root structure will hold soil in place and prevent erosion. This benefit also extends to wild cannabis; there are anecdotal reports of farmers routinely allowing feral populations of cannabis to grow in irrigation ditches to help maintain their structural integrity. After harvest, cannabis's root systems play a vital role in decompacting the soil, allowing for better flow of water and nutrients through the soil. Farmers harvest the plant at the stem, leaving the roots in the ground. The roots will decompose in the soil over the winter, leaving channels of space for air and water to flow through, and the decaying plant matter will provide beneficial nutrients to the soil.

PHYTOREMEDIATION AND ABSORBENCY

A curious property of industrial hemp is its ability to absorb soil contaminants, such as heavy metals, pesticides, and crude oil, without dying. This quality is not unique to industrial hemp—many plant species have this ability, but since hemp is such a prolific biomass accumulator researchers suggest it may be an ideal candidate for a practice called phytoremediation. Phytoremediation is the practice of intentionally cultivating specific plant species to remove contaminants from the water or soil. This is not without its risks, as I will mention below and also discuss in the context of quality assurance in chapter 11.

Repeated studies show that *Cannabis sativa* has an extreme affinity to take up heavy metals from soil with minimal detrimental effects to its growth. In fact, the Ukrainian government successfully planted hemp in the Chernobyl exclusion zone to try to lower cadmium concentrations in the soil. Additionally, studies show hemp's ability to take up copper, zinc, magnesium, chromium, lead, nickel, selenium, and cadmium from a variety of soil environments (Citterio et al. 2003, 243; Stonehous et al. 2020, 4221). Since humans require small amounts of dietary selenium for health, hemp could help lower soil selenium concentrations (in the absence of other contaminants) and provide nutritional supplements in one crop. Many phytoremediative crops do not grow as well or as quickly as hemp in variable environments, limiting their utility. This environmental adaptability makes hemp more attractive than other crops as a phytoremediator; furthermore, hemp grows densely and quickly, improving its phytoremediative performance relative to slower growing plants like trees.

Cannabis has also been proposed as a phytoremediator for hazardous waste sites. Superfund sites, brownfield development sites, and even heavily polluted aquatic ecosystems could all benefit. Currently, there is a research project in Lake Okachobee, FL where a private-public partnership is looking to remove nitrogen and phosphorus pollution. Heavy agricultural activity in central Florida has left Lake Okachobee heavily polluted from agricultural runoff. The heavy concentrations of nitrogen and phosphorus in the lake lead to heavy algal blooms in nearby rivers and marshes and are the leading cause of red tide in Florida. Solving this problem would confer not only an environmental benefit, but also an economic one. The Florida Department of Health estimates red tide costs the state on average $22 million per year in health costs, as well as additional lost revenues for local businesses, and decreased catch for local fisheries (Florida Department of Health 2012).

Any cannabis grown for phytoremediative purposes should not be used for human consumption. If you are sourcing any cannabis material for use in products destined for human consumption, be sure you understand your supply chain and how producers should operate and test for contaminants in harvested plant matter (see also chapter 11). Cannabis contaminated in this way obviously eliminates a potential revenue stream; however, harmful soil contaminants accumulate in the highest concentrations in the leaves of the plant, which have the least economic value (Linger et al. 2002, 39). The parts of the plant with high concentrations of contaminants can be sent off for further processing and disposal, whereas useful parts of the plant like fiber and seeds can be harvested and turned into biofuels or building materials. Considering the level of environmental pollution around the globe, cannabis presents a compelling possibility for phytoremediation while simultaneously harvesting useful products.

5

BIOCHEMISTRY OF CANNABINOIDS

Thinking back to our theoretical group of early *Homo sapiens* and their first encounter with the psychoactive properties of *Cannabis sativa*, it is amazing how much we now understand about cannabis's glandular trichomes and the cannabinoids they secrete. Those first experiences of early humans, whenever they happened, must have been so transcendental that it set in motion thousands of years of discovery into the underlying mechanisms of what makes us feel high.

We now know that the cannabinoid compounds secreted by the glandular trichomes covering the female plant's flowers and foliage are what lie behind the psychoactive effects of cannabis. The cannabinoids in the form they are created within the living plant do not get humans high but become psychoactive when they are decarboxylated through heat or time (we will explain what that means later). When we smoke or ingest cannabinoids derived from marijuana, we feel high. This chapter will cover the biochemistry of how *Cannabis sativa* produces cannabinoids and give a brief overview of the various physiological responses that humans have to these cannabinoids. Chapters 6 and 7 will focus on strategies to effectively solubilize cannabinoids into beer and beverages. The glandular trichomes of cannabis, like all such trichomes in plants, also secrete a multitude of aromatic hydrocarbons. These mostly consist of terpenes, which we will talk about in chapter 8.

Generally, there is a complex mixture of cannabinoids and terpenes in any given cannabis plant. All cannabis plants are *Cannabis sativa*, this is true, but there are multiple varieties, cultivars, and "strains" within the species that have arisen over centuries and millennia thanks to humans breeding cannabis for specific uses. Given the confusing taxonomic nomenclature that exists around hemp and marijuana types (see chapter 3), especially the naming of strains in the marijuana market, academic researchers have increasingly turned to analyzing the biochemical profile of cannabis plants to better identify them unambiguously. These "chemical varieties" are usually referred to as chemotypes or **chemovars**. Genetics within individual *Cannabis* chemovars do play a large role in the overall cannabinoid makeup, but additional environmental factors complicate matters, for reasons we will get into later in the chapter. Thus, the chemovar concept, by taking the overall biochemical attributes of the plant as a key identifier, help discriminate between plants that may otherwise be morphologically and/or genetically indistinguishable.

WHAT ARE CANNABINOIDS?

Cannabinoids are a group of terpenophenolic[1] compounds and their derivative by-products present mainly in *Cannabis sativa*. Cannabinoids bind to and activate receptor molecules in the mammalian nervous system. The presence of these cannabinoid receptors was discovered decades after the structures of cannabidiol (CBD) and delta-9 tetrahydrocannabinol (THC) were elucidated; shortly thereafter, scientists discovered that mammals produce their own molecules that activate cannabinoid receptors and called these endocannabinoids. Humans and other animals naturally produce endocannabinoids as part of the regulation and expression of the endocannabinoid system, which is responsible for a wide variety of functions essential to the immune, circulatory, and central nervous systems

[1] We first encountered terpenes in chapter 4 (p. 91). A terpenophenolic compound consists of a terpene molecule combined with a molecule that contains a phenol group (C_6H_5OH), which is essentially a six-sided ring of carbon atoms with an extra oxygen and hydrogen attached.

(Rodriguez de Fonseca et al. 2005, 3). Further, other plants have been found to synthesize molecules that also interact with cannabinoid receptors; this has led to the broader classification of such compounds as "phytocannabinoids," that is, any plant-derived compound that interacts with cannabinoid receptors in mammals and that may or may not be chemically similar to *C. sativa* cannabinoids (Small 2017, 220). Note the "or may not" part of that definition. Phytocannabinoids and endocannabinoids can vary widely in chemical structure and pharmacological or biological activity; thus, to use these terms interchangeably with the "classical" cannabinoids derived from *C. sativa* is misleading. It's helpful to understand that all these compounds elicit a biochemical or physiological response in the endocannabinoid system; however, our definition of "cannabinoid" moving forward will only include terpenophenolic compounds produced by *C. sativa*.

Almost 150 cannabinoids have been identified in cannabis, with more discovered every year (Hanuš et al. 2016, 1388). Although most of these cannabinoids are naturally present in the plant at low concentrations, modern chemistry and advances in the fields of genomics and plant biology have allowed researchers to identify many novel cannabinoids, and even synthesize them for further study.

The two cannabinoids predominant in cannabis are THC and CBD. Broadly, the highest concentrations of THC are found in marijuana biotypes and the highest concentrations of CBD are found in hemp biotypes. There are exceptions to this paradigm but, generally, it is useful to think of cannabinoid production in this context.

ROLE OF CANNABINOIDS IN THE PLANT

While there are certain evolutionary advantages that cannabinoid excretion confers on the cannabis plant, humans have played a large role in its current genetic and biochemical makeup. As with many plants that are domesticated today, humans bred cannabis for specific purposes—in this case to maximize glandular trichomes and, therefore, cannabinoid production—so what was once true about the evolutionary advantages cannabinoids provide in the wild bear only a passing resemblance to what they provide for the plant today.

Most researchers theorize that cannabinoids are a part of the plant's natural defense system, designed to ward off predators and disease. Cannabinoids as they are naturally synthesized by the plant are a type of carboxylic acid and, during stages of growth, this acidic state maintains a protective function.

Studies have shown that the cannabinoids cannabigerolic acid (CBGA) and tetrahydrocannabinolic acid (THCA) induced cell death in 100 percent of cells in tissue cultures of various plants and insects. Researchers further confirmed the cannabinoids were responsible for this effect by comparing them with olivetolic acid, which is a precursor molecule used in the biosynthesis of CBGA and THCA. Olivetolic acid did not have any effect on the cells in the tissue cultures studied, whereas CBGA and THCA exerted their effects within ten days (Hazekamp et al. 2010, 1060).

Whereas cannabinoids naturally present in the plant are in an acidic form, it is important to remember that humans process the harvested plant matter to bring about the decarboxylation of cannabinoids, because that is what makes them more bioactive when ingested. We will come back this once we have reviewed the general mechanisms of cannabinoid biosynthesis.

CANNABINOID BIOSYNTHESIS

Before we jump into the specific mechanisms of cannabinoid biosynthesis, it's important to state that terpenophenolics are not unique to *Cannabis sativa*. For one thing, the hop plant (*Humulus lupulus*) synthesizes humulone (alpha acid) and lupulone (beta acid), two terpenophenolics of great interest to brewers. Maybe this does not come as much of a surprise to those who know *Cannabis* and *Humulus* are related genera of the Cannabaceae family. While there are limited occurrences of terpenophenolics in other plant species, these compounds are best-known as signature secondary metabolites of Cannabaceae, with cannabinoids being indicative of *C. sativa*.

The biosynthetic pathways for terpenophenolics in hop species (bitter acids and xanthohumol pathways) and cannabis (cannabinoid pathway) have several commonalities, with a similar process of polyketide formation, aromatic prenylation, and cyclization. The diversity of terpenophenolics in the Cannabaceae family arises from variations on this pattern (Page and Nagel 2006, 180). Cannabinoid biosynthesis begins with polyketide formation, which results in the formation of olivetolic acid. Olivetolic acid is produced by fatty acid synthesis, where a starting molecule of hexanoyl-coenzyme A undergoes three rounds of extension with a malonyl-coenzyme A unit, each round involving an enzymatic condensation reaction that joins the hexanoyl primer with a malonyl extender. The resulting tetraketide-coenzyme A is a linear molecule, which must then undergo several coordinated steps involving a polyketide synthase enzyme and a polyketide cyclase enzyme, the latter called olivetolic acid cyclase (Andre, Hausman, and Guerriero 2016, 3). Olivetolic acid is one of the precursors necessary for cannabinoid biosynthesis; it must be joined to another precursor, geranyl pyrophosphate, to create a cannabinoid.

Geranyl pyrophosphate (a.k.a. geranyl diphosphate) derives from the methylerythritol phosphate (MEP) pathway, sometimes called the deoxyxylulose phosphate (DOXP) pathway. The MEP pathway is a biosynthetic pathway common to plants and some single-celled organisms and is an important pathway for producing compounds necessary for cell membrane maintenance, binding intracellular proteins, and synthesizing hormones. The pathway is also responsible for the biosynthesis of terpenes.

Cannabinoid synthesis now proceeds via aromatic prenylation, whereby geranyl pyrophosphate is joined with olivetolic acid, a reaction catalyzed by the enzyme geranylpyrophosphate:olivetolate-geranyltransferase (Hazekamp et al. 2010, 1041). The people that name enzymes certainly have an improvisational sensibility when it comes to naming conventions. This prenylation forms the first cannabinoid, cannabigerolic acid (CBGA), which is the central precursor from which all other cannabinoids are made. Cannabigerol (CBG), which is the decarboxylated form of CBGA, was first discovered by Israeli scientists in 1964; at the time these scientists thought they had discovered the cannabinoid responsible for marijuana's psychoactivity. It wasn't until later, with the discovery of THCA, that researchers realized CBGA was the "mother" cannabinoid to all other cannabinoids.

The final phase of the cannabinoid synthesis process is cyclization. This involves the prenyl chain (the bit derived from geranyl pyrophosphate) of CBGA undergoing one of several possible enzymatic cyclization reactions, forming the three main derivatives of CBGA: delta-9 tetrahydrocannabinolic acid (THCA), cannabidiolic acid (CBDA), and cannabichromenic acid (CBCA). The oxidocyclase enzymes responsible for making these compounds are THCA synthase, CBDA synthase, and CBCA synthase, respectively (Andre, Hausman, and Guerriero 2016, 3). Genetics plays a large role during this stage of the biosynthetic process. All cannabis chemovars follow the same pathway to synthesize CBGA, but now some divergence is seen, as different chemovars produce varying amounts of the synthase enzymes. Any given cannabis chemovar will usually produce at least some of all three of these enzymes, but some chemovars will produce more THCA synthase than CBDA synthase, or vice versa; some chemovars will produce predominantly CBCA, though it is rare.

Recent research has worked to identify and isolate curious examples of rare and exotic cannabinoids. A great example is the discovery of a compound called delta-9 tetrahydrocannabivarin, or THCV. Marijuana smokers purport that THCV suppresses appetite and gives a more pleasant high. As its name suggests, THCV is similar in structure to THC; however, while the reaction mechanisms involved in its biosynthesis are the same (i.e., polyketide formation, aromatic prenylation, and cyclization), instead of the enzymatic prenylation of olivetolic acid, geranyl pyrophosphate condenses with divarinolic acid, creating cannabigerovarinic acid CBGVA (Hillig and Mahlberg 2004, 966; Hazekamp et al. 2010, 1043). CBGVA is then cyclized into delta-9 tetrahydrocannabivarinic acid (THCVA), cannabidivarinic acid (CBDVA), and cannabichrovarinic acid (CBCVA). Variations like this continue to be discovered and may potentially reveal highly novel and beneficial cannabinoids for human use.

With the creation of some combination of THCA, CBDA, CBCA, and a small amount of unconverted CBGA, the enzymatic portion of cannabinoid biosynthesis is complete. It is around this time that the plant is harvested and dried. Through further processing, more cannabinoids can be created.

NON-ENZYMATIC SYNTHESIS OF DIFFERENT CANNABINOIDS

We've established that cannabinoids biosynthesized in the plant exist predominantly in an acidic form, although the plant may produce low levels of neutral (non-acidic) cannabinoids. What do we mean when we say the cannabinoids are acidic? While many different molecules can act as acids, with cannabinoids what makes them acidic is the presence of a carboxyl group, which is a specific arrangement of atoms consisting of a carbon, two oxygens, and a hydrogen (often represented as COOH or C(=O)OH in chemical formulas). When part of a larger molecule, the carboxyl group will tend to release its hydrogen very easily; thus, the presence of the carboxyl means the compound as a whole acts as an acid.

It is primarily human or environmental interventions that transform these acidic cannabinoids into a neutral state. The transformation process is called **decarboxylation**, which means to lose a carboxyl group; when this happens, the carboxyl is replaced by a single hydrogen, with a net loss of carbon dioxide (CO_2). Most easily accomplished with heat, decarboxylation can also occur over time with exposure to oxygen or ultraviolet light. Decarboxylation is the mechanism that turns the acidic THCA naturally produced in the plant into the neutral (decarboxylated) psychoactive THC that humans ingest. The obvious way to understand this is to think about the most common way of consuming marijuana: smoking. A marijuana joint of ground up inflorescences contains mostly THCA. When the joint burns, the heat rapidly decarboxylates the THCA into THC.

The final form of cannabinoids are **artifact** cannabinoids. Artifact cannabinoids form under storage conditions where some combination of heat, oxygen, or ultraviolet light catalyzes the degradation of the precursor cannabinoid. Artifact cannabinoids exist in both acidic and neutral forms, depending on whether the precursor cannabinoid was decarboxylated. A great example comes from an early study of cannabis chemistry, when the chemist Thomas Wood claimed discovery of the psychoactive compound in marijuana that makes users high (Wood, Newton Spivey, and Easterfield 1899, 20). Wood extracted and isolated this newfound cannabinoid and called it cannabinol (CBN). However, because in the 1890s Wood likely required his marijuana sample to be shipped over long distances, the combined storage stress and time probably converted most of the THC into CBN. This gave Wood and his partners the mistaken impression that marijuana produces CBN, whereas we now know CBN is an artifact cannabinoid of THC. All three primary biosynthesized cannabinoids— THCA, CBDA, and CBCA—produce artifacts, and some produce more than one.

Identifying and synthesizing all the variants of the ten main cannabinoid groups (see sidebar, p. 116) requires sophisticated analytical chemistry techniques. Isolating individual cannabinoids for analysis requires advanced separation techniques, most often some combination of gas chromatography (GC) combined with mass spectroscopy (MS) or flame ionization detection (FID); alternatively, depending on the target cannabinoid, high-performance liquid chromatography (HPLC) is used. Due to the complexity of cannabis's constituents, accurately quantifying every component using GC can be challenging. It is possible to get false or inaccurate readings of cannabinoids in any given sample—conditions such as poor sample preparation, thermal degradation, column activity, and calibration protocols can all present challenges. It is best to employ a variety of techniques to get accurate and reproducible quantification. In recent years, some laboratories using nuclear magnetic resonance (NMR) spectroscopy claim to have higher precision when measuring cannabinoid content. The world of cannabinoid testing can be variable, which is worth mentioning considering the implications for failing to get repeatable, reliable quantification of cannabinoids. We will go into further detail on this topic when we look at quality assurance in chapter 11.

Cannabis Cannabinoids Classified by Structure

Cannabinoids are classified into 10 main structural groups (Brenneisen 2007, 18–25). Each structural classification applies to both acidic and neutral states; they are labeled here under neutral states for conciseness.

Cannabigerol Type (CBG): Seven variants within the CBG group. Can by enzymatically cyclized into THC, CBD, and CBC.

Delta-9 Tetrahydrocannabinol Type (THC, or Δ^9-THC): Nine variants within the THC group. Can be isomerized into delta-8 THC (a THC artifact), which, over time, will degrade into the artifact cannabinol (CBN) with exposure to heat, oxygen, or ultraviolet light.

Cannabidiol Type (CBD): Seven variants within the group. Will degrade over time into its artifacts cannabielsoin (CBE) and cannabinodiol (CBND) with exposure to heat, oxygen, or ultraviolet light.

Cannabichromene Type (CBC): Five variants within the group. Will degrade over time into its artifact cannabicyclol (CBL) with exposure to heat, oxygen, or ultraviolet light.

Cannabinol Type (CBN): Seven variants within the group; artifact of THC.

Delta-8 Tetrahydrocannabinol Type (Δ^8-THC): Two variants within the group; Δ^8-THC is a heat-generated artifact of THC.

Cannabielsoin Type (CBE): Five variants within the group; artifact formed from CBD.

Cannabinodiol Type (CBND): Two variants within the group; artifact formed from CBD.

Cannabicyclol Type (CBL): Three variants within the group; heat-generated artifact from CBC.

Cannabitriol Type (CBT): Nine variants within the group.

Miscellaneous types: Includes rare and unclassified cannabinoids. Some cannabinoids have only been discovered through synthetic chemistry and their relation to the "classical" *Cannabis* cannabinoids and derivatives is unclear.

CANNABINOID BIOSYNTHESIS FROM GENETICALLY MODIFIED ORGANISMS

With modern gene editing techniques, it is now possible to produce cannabinoids without ever growing a cannabis plant. One such instance is editing the genome of yeast (*Saccharomyces cerevisiae*) to code for the enzymes from the primary cannabinoid biosynthetic pathways in *Cannabis sativa* that are responsible for olivetolic acid and cannabinoid synthesis (Luo et al. 2019, 123–124). Following a series of genetic editing techniques, researchers then showed that the modified yeast could create the primary cannabinoids CBGA, CBDA, THCA, THCVA, and CBDVA from a simple sugar substrate (Luo et al. 2019, 124–125).

This development presents one of the most compelling opportunities for creating high-quality cannabinoids without ever having to disturb the natural environment with modern farming practices. The genetically modified yeast could create cannabinoids in days, utilizing rudimentary processes and, theoretically, capable of producing the world's cannabinoid demand in a single location. These findings were only published in 2019 so the technology is still in its infancy, but it will be an exciting development to watch as it is scaled and perfected.

OVERVIEW OF CANNABINOID GROUPS

Cannabigerol (CBG)

Cannabigerol (CBG), or more precisely cannabigerolic acid (CBGA), is the precursor compound to all other cannabinoids. CBGA has shown considerable antibacterial activity against gram-positive bacteria (ElSohly and Slade 2005, 542) and both CBGA and CBG have shown antifungal properties (Russo and Marcu 2017, 81). Normally, CBGA appears in relatively low concentrations in harvested and dried cannabis; however, researchers have bred chemovars that lack the synthase enzymes to cyclize CBGA into its downstream derivatives, effectively halting the usual biosynthetic pathway at this stage and yielding 100 percent CBGA in harvested inflorescences (de Meijer and Hammond 2005, 190). CBG has no demonstrated intoxicating effects and behaves similarly to CBD in the body. As well as antibacterial and antifungal activity, studies of CBG have demonstrated antioxidant and anti-inflammatory properties. Additionally, researchers theorize that CBG could be an effective treatment for relieving intraocular pressure as a therapy for glaucoma (Russo and Marcu 2017, 81). CBG—along with CBC, THC, CBD, and CBN—has shown to have potent activity against MRSA and could potentially lead to future therapies to treat bacterial infections (Appendino et al. 2008, 1427). There are currently no FDA-approved therapies utilizing CBG.

Delta-9 Tetrahydrocannabinol (THC)

Delta-9 tetrahydrocannabinol (THC, or Δ^9-THC) garners the most attention in cannabinoid chemistry because of its pharmacological and psychoactive properties. It is a controlled substance globally, and it is illegal to possess, sell, or study it in most nations. Currently, THC is legal to possess only in Canada, Georgia, South Africa, and Uruguay. There are, of course, many exceptions to THC bans, especially when it is used for medical purposes under professional supervision. Legalization of THC for recreational use has made significant progress over the last ten years, and will no doubt expand in the future.

The most notable human physiological response to THC is the characteristic "high." This can vary between users, depending on such parameters as dosage, personal tolerance, and environment. THC is technically hallucinogenic, meaning it alters the perception of reality—especially when taken in high doses—by affecting how neurons in the brain communicate with each other. Mental responses include impaired reaction time, memory retention, and judgement; feelings of euphoria; heightened sensory awareness, creativity, and empathy; altered sense of time and space; and enhanced appetite, introspection, and sexual desire (Zablocki et al. 1991, 78). These mental responses make THC, and marijuana generally, one of the most popular recreational drugs on the planet. The World Health Organization estimates that roughly 2.5 percent of the global population—some 200 million people—consumed marijuana sometime in the last year.[2] Common physical responses include dry and red eyes, reduced intraocular pressure, dry mouth, muscle relaxation, and raised heart rate (Wardle, Marcus, and de Wit 2015, 16).

These physiological effects help inform the potential pharmacological efficacy of THC for certain conditions. In some cases, THC could be considered a beneficial therapeutic for people with bipolar disorder (Grinspoon and Balakar 1997, 142) and other mental disorders, but often the psychoactive effects are a barrier to THC being adopted as a medicine. Most researchers view the psychoactive effects as a tradeoff to getting an efficacious result, but many others seek to understand the underlying mechanisms in the hopes that one day they can "turn off" the psychoactive effects while keeping the beneficial therapeutic effects. To complicate matters, THC has biphasic effects, meaning it can produce one outcome at one dosage and produce the opposite outcome at a different dosage.

---

2 "Cannabis Facts and Figures," World Health Organization, accessed April 28, 2020, https://www.who.int/substance_abuse/facts/cannabis/en/.

Nevertheless, globally, THC use as a medicine is at an all-time high. Studies show that THC can increase appetite in patients with nausea or eating disorders and improve a patient's mood (Russo 2011, 1354). Additionally, THC has been shown to be a powerful anti-inflammatory agent and pain reliever, showing as much as twenty times the anti-inflammatory power as aspirin and twice the power of hydrocortisone (Evans 1991, 66). As a pain reliever, THC works differently than more traditional pain medications like opioids, which bind to pain receptor sites in the brain and block the reception of pain. THC reduces the perception of pain by modulating pain pathways. The two THC drugs approved by the US Food and Drug Administration (FDA) are a synthetically pure THC application called dronabinol (marketed under the name Marinol®) and a blend of plant-extracted THC and CBD (called Sativex®). Both drugs are approved to treat symptoms of nausea and vomiting, as well as to stimulate appetite for patients with serious diseases such as cancer, HIV/AIDS, and multiple sclerosis (Russo and Marcu 2017, 80).

While THC does have demonstrated pharmacological benefits, there are downsides to its use for both individuals and society. Acute health effects of using cannabis include impaired cognitive development, especially among minors; lack of free recall of information; impaired coordination and motor function; inattention; and risk of impairment while operating a vehicle (WHO 2018c, 59). Chronic use of THC can lead to impaired cognitive function, including lack of organization and integration of complex information into thought; dependence; exacerbation of schizophrenia symptoms for people susceptible to developing schizophrenia; impaired lung function and bronchitis; and risks to fetal development (when used during pregnancy). While most experts agree that the risks of serious personal and societal harm from widespread THC use is low—especially when compared with other legal intoxicants like alcohol—they are not absent.

Cannabidiol (CBD)

Cannabidiol (CBD) was first identified in the 1940s and is the primary cannabinoid produced by industrial hemp. It can also be found in high concentrations in some marijuana strains, although this is less common, likely due to humans selecting for high THC potency. This is of course a modern development—for most of history there was no need to breed cannabis to have artificially low or high THC levels and, in some cases, artificially high CBD levels.

CBD is non-intoxicating and has been extensively studied for many pharmacological uses. CBDA has considerable antimicrobial properties and has shown promise in treating many types of infections, including methicillin-resistant *Staphylococcus aureus*, better known as MRSA (Petri 1988, 341; Appendino et al. 2008, 1428). CBD has powerful antioxidant properties and purportedly is a more powerful antioxidant than both vitamins C and E (Hazekamp et al. 2010, 1057). The only FDA-approved CBD-based drug, called Epidiolex®, is prescribed for the treatment of epileptic seizures associated with Lennox-Gastaut syndrome and Dravet syndrome. While this is the only approved CBD therapy, there is an impressive range of potential therapies for which CBD may be suitable. Laboratory studies and clinical trials have demonstrated that CBD has powerful anti-inflammatory, antipsychotic, and neuroprotective properties in addition to its anticonvulsive and antioxidant properties. Active studies suggest CBD shows promise in treating a number of neurological disorders, including Parkinson's disease, Huntington's disease, Alzheimer's disease, multiple sclerosis, and amyotrophic lateral sclerosis (Russo and Marcu 2017, 80), although it should be said that there are no proven CBD-based therapies for these diseases to date.

Many companies sell CBD nutritional supplements, and consumers take these products for the purported benefits of, among others, pain and inflammation reduction, relaxation, sleep improvement, and reduced anxiety. These benefits are not supported by the FDA, and it is illegal for companies to market their products using definitive claims that consumers will reap these benefits. Nevertheless, US consumers bought $800 million of CBD products in 2019 and the market continues to grow.

Some studies suggest that a dose of up to 400 mg CBD has no adverse side effects; however, such findings are not conclusive. When taken in high doses, CBD has been shown to cause hepatoxicity (liver damage), which, in fact, is listed as a warning on Epidiolex packaging. This effect starts to appear at a dose of 20 mg/kg body mass, a dosage ranging from 1000 mg to 1600 mg per day (Ewing et al. 2019, 11).

The World Health Organization concluded in 2018 that CBD has yet to show abuse potential in clinical studies. In its *Cannabidiol (CBD): Critical Review Report*, the WHO states that, "To date, there is no evidence of recreational use of CBD or any public health-related problems associated with the use of pure CBD" (WHO 2018a, 5). The report also states that "in an animal drug discrimination model CBD failed to substitute for THC."

That is not to say that CBD does not get used in conjunction with recreational marijuana. Studies show that CBD inhibits the metabolism of THC by blocking the conversion of THC into more psychoactive compounds in the body; this effect is even mentioned in the patent for Sativex (Whittle 2003).

Many commercial marijuana products use CBD in their marketing by expressing ratios of CBD to THC to differentiate some of their products and the intended high. Studies do show that CBD can specifically counteract some of the adverse effects of consuming too much THC, such as increased heart rate and anxiety (Russo and Guy 2006, 234)

Cannabichromene (CBC)

Cannabichromene (CBC) is the third primary cannabinoid to be enzymatically isomerized from CBG during cannabinoid biosynthesis, along with CBD and THC. Normally a minor constituent of cannabis in both marijuana and industrial hemp varieties, even those varieties considered high in CBC rarely contain more than one percent CBC by weight. Breeders have been able to increase CBC content by selecting for recessive genes that boost the expression of CBC synthase enzymes (de Meijer 2009). CBC or CBC-like derivatives have also been found in *Rhododendron* species; however, these are not classed as scheduled drugs by the DEA (Iwata and Kitanaka 2011, 1409). CBC is non-intoxicating, much like CBD. Studies show strong anti-inflammatory properties in animal studies (DeLong et al. 2010, 131), but no therapies using CBC are currently approved by the FDA.

Cannabinol (CBN)

Cannabinol (CBN) is an artifact cannabinoid, being an oxidative by-product of THC. Heat, oxygen, and ultraviolet light exposure can all catalyze the conversion of THC to CBN, although the conversion does not completely transform all THC to CBN, instead transforming it into both CBN and Δ^8-THC (ElSohly and Slade 2005). Researchers have found that CBN is mildly intoxicating, but far less so than THC, with one study claiming it is up to four times less potent than THC (Izzo et al. 2009, 516). Non-clinical and laboratory studies show that CBN has sedative, anticonvulsant, anti-inflammatory and antibiotic properties, including treating MRSA in a laboratory study (Russo and Marcu 2017, 83; Appendino et al. 2008, 1427). There are currently no FDA-approved products or therapies utilizing CBN.

Delta-8 Tetrahydrocannabinol (Δ^8-THC)

Delta-8 tetrahydrocannabinol (Δ^8-THC) is an artifact cannabinoid derived from the thermal degradation of THC. The thermal degradation of THC is not efficient for creating Δ^8-THC, and some THC is converted to CBN via oxidative aromatization. The configuration of Δ^8-THC is far more thermodynamically stable than THC and will not degrade into CBN. Δ^8-THC is roughly 20 percent less intoxicating than THC (Brenneisen 2007, 25). Clinical trials have demonstrated Δ^8-THC has antiemetic, anxiolytic, analgesic, neuroprotective, and appetite-stimulating properties. One Israeli study found that Δ^8-THC could be safely administered in much higher doses than THC in pediatric cancer patients and, as such, could be more effective in treating nausea and vomiting while minimizing side effects associated with other antiemetic drugs (Abrahamov, Abrahamov, and Mechoulam 1995, 2097). More recently, marijuana companies have been marketing pure isolates of Δ^8-THC as a lower-potency alternative to products containing THC. These companies tout the "relaxation and clear-headedness" that differs from physiological responses to traditional marijuana-based THC products. There have been several clinical trials to assess the medical efficacy of Δ^8-THC and surely there will be more. Currently, there are no Δ^8-THC-based products or therapies approved by the FDA, and Δ^8-THC remains a controlled substance in the US and in general internationally.

Cannabielsoin (CBE)

Cannabielsoin (CBE) is an artifact cannabinoid resulting from thermal, photochemical, or oxidative degradation of CBD. CBE is considered non-psychoactive, although we currently know very little about its biological or pharmacological effects. There are no products or therapies utilizing CBE approved by the FDA.

Cannabinodiol (CBND)

Cannabinodiol (CBND) is an artifact cannabinoid derived from the thermal, photochemical, or oxidative degradation of CBD. CBND is mildly intoxicating, making it something of an oddity since it is an artifact cannabinoid derived from a non-psychoactive precursor. Early researchers also theorized that CBND could derive from the photochemical conversion of CBN; however, those findings have since been questioned (ElSohly and Slade 2005, 544). Little is currently known about this cannabinoid. There are no products or therapies utilizing CBND approved by the FDA.

Cannabicyclol (CBL)

Cannabicyclol (CBL) is an artifact cannabinoid resulting from the photochemical action of ultraviolet light on CBC. Since it is produced in such small quantities (along with its precursor, CBC), little is known about CBL's pharmacological effects. CBL is considered non-psychoactive but has not been demonstrably proven so. Unlike many of the other cannabinoid groups, CBL is not a scheduled substance by US or international conventions. There are no products or therapies utilizing CBL approved by the FDA.

Cannabitriol (CBT)

Cannabitriol (CBT) is the only major cannabinoid group with an unknown biosynthetic pathway in cannabis. It naturally occurs in small quantities, making its isolation and study complicated. Little has been studied about the biological, psychoactive, or pharmacological effects of CBT. There are no products or therapies utilizing CBT approved by the FDA.

SOLUBILIZING SHELF-STABLE CANNABINOIDS IN BEVERAGES

While this is a book on cannabis and beer, I will use the loose term of "beverage" in this chapter to discuss the theory of how to create shelf-stable cannabis beverages. Chapter 7 addresses dosing cannabinoids into beer specifically; for practical and legal purposes, that chapter concerns CBD exclusively. I address beverages in general here because beer is primarily water based and, with respect to the disposition of lipophilic molecules, behaves similarly to non-alcoholic beverages. While ethanol can be an excellent medium for solubilizing cannabinoids, not every beverage maker can utilize ethanol in their formulations. Additionally, the ability to solubilize cannabinoids is relatively proportional to ethanol concentration. Cannabinoids will readily dissolve into high-proof spirits and be physically stable; in relatively low-alcoholic beers, the ethanol content is insufficient to act as a delivery medium for creating shelf-stable cannabinoid-based beverages.

This chapter describes the cannabinoid beverage formulation process from the point of view of a cannabinoid manufacturer rather than the point of view of a brewer, because brewers are typically ill-equipped to do this work. Producing cannabinoid concentrates is technically challenging and will consume significant administrative and regulatory resources that few brewers have. Additionally, this chapter will focus more on the needs of a commercial brewer/beverage maker. For those of you at home, there are more low-tech options that we will discuss in chapter 7 (p. 153). For commercial manufacturers dosage accuracy is critical, not only for regulatory compliance but also for providing reliable and repeatable experiences for your customer base. At time of writing, the method outlined below has proven to be the most reliable method for dosing the correct cannabinoid content without negatively affecting the presentation and organoleptic qualities of your beverage.

The fundamental challenge with getting cannabinoids (and, to a lesser extent, terpenes) into a beverage is that they are lipids (fats) and do not solubilize in water. Granted, with a high enough inclusion rate of raw cannabis, one could solubilize a measurable quantity of cannabinoids into a beverage, but it would be financially and logistically prohibitive to do so. Even then, assuming we could keep them perfectly homogenous in a cold bright tank, the cannabinoids would likely separate back out of the aqueous phase, leaving an unappealing slick of oily resin on the beverage. One study in the Netherlands showed that preparing a basic marijuana tea by boiling marijuana flower in water for thirty minutes dissolved only 17 percent of the available THC into the beverage. After being stored for one day, the THC had fully precipitated out of the tea (Hazekamp et al. 2007, 85).

The solution to this problem is to create an emulsion, a relatively stable mixture of immiscible liquids such as oil and water. Emulsions exist in many foods and beverages, for example, oil and vinegar salad dressings and mayonnaise. While home cooks can emulsify a salad dressing by shaking it, the dressing will soon separate. Hence, food technologists develop sophisticated methods that allow emulsions to be stable in the long term. Shipping and storage conditions, total shelf life, and environmental stressors like pH influence how emulsions behave and make emulsion technology an interesting, cutting-edge field of study.

Cannabinoid emulsion systems are known as oil-in-water emulsions, with cannabinoids being the oil component dispersed into a water base. Emulsions can also be formulated to carry water-based compounds in oils and, unsurprisingly, are known as water-in-oil emulsions (one example is mayonnaise); however, we will not discuss water-in-oil emulsions any further. The general formula for a cannabinoid emulsion system consists of the bioactive compound (cannabinoids) dissolved in a food-grade carrier (oil), which is then encapsulated by a hydrophilic surfactant (emulsifier). Perhaps you may have heard of "water-soluble cannabinoids," which cannabis companies use to claim that their products can defy the laws of physics. This is a misnomer: the "solubility" comes from the hydrophilic properties of an emulsifier that reduces the surface tension and enables the beverage and the encapsulated oil droplet to interact.

There is a wide array of methods employed in emulsification technologies, with many ways to achieve success. Individual beverages must be evaluated on the basis of their physical and chemical properties and then matched to a suitable emulsion system. If you are exploring options, I suggest reaching out to multiple vendors to see how their systems differ. The most important component of any emulsion system is that it be GRAS—generally recognized as safe—by your country's national food regulatory body. Any reputable company creating cannabinoid emulsions will have supporting paperwork to prove that their emulsion system qualifies under GRAS guidelines, and that their processing conditions do not alter the chemical structure of the emulsifier, which could potentially create a new compound of unknown toxicity. Look for vendors with a well-trained technical staff to help you through the trial-and-error process of figuring out what works best for your product.

None of these technologies are novel to the cannabis industry. Food researchers and chemical engineers designed many of these technologies decades ago to solve the myriad problems relating to stabilizing and shipping food and beverages around the globe. Many of these technologies are commonly applied throughout many industries and products, including dairy, fresh and concentrated fruit juices, soft drink flavorings, and fragrances, as well as pharmaceuticals and nutraceuticals. The interesting component is that while most of the technologies came from the food and beverage industry, most of the focus on how to improve these technologies for the cannabis industry is informed by research from the pharmaceutical industry. This makes sense; cannabis exists in a space somewhere between a recreational intoxicant—often in the form of food and beverages—and as a therapeutic.

Lastly, most emulsion systems are proprietary to companies. The aim of this chapter, and of this book in general, is to arm the reader with quality information so that they can ask informed questions about individual technologies. While I have personally worked with many different emulsion systems, nothing in this chapter reveals any individual company's protected intellectual property.

FORMULATING A CANNABINOID-INFUSED BEVERAGE

Creating a cannabinoid-infused beverage follows a similar process to creating other cannabinoid-infused edibles or concentrates. While there are minor technical differences between processing CBD and THC for beverage applications, the overall concept is the same for both.

The basic principle involves creating a pure, stable, decarboxylated cannabinoid concentrate and then encapsulating it in an emulsifier that allows the cannabinoid concentrate to be dosed into a water-based beverage. One technology in the process that should be familiar to commercial brewers is liquid carbon dioxide extraction, which can be found applied in any major hop-growing region. What is produced from a liquid carbon dioxide extractor looks similar for both hops and cannabis; it's useful to hold that analogy in your head when understanding the principles behind creating cannabinoid beverages. One point where the analogy diverges is that cannabinoid extracts are added cold to the beverage and will not have the aid of heat to help with solubilization.

Step 1: Biomass Preparation

The process of formulating a cannabinoid beverage begins by grinding dried, cannabinoid-rich cannabis biomass. It is important to dry the biomass down to at least 12–15 percent moisture content, otherwise the subsequent extraction step will have to remove the water from the concentrate later, adding both complication and expense. The drying step is also commonly the time for decarboxylation to occur, which transforms naturally acidic cannabinoids into their neutral counterparts that are pharmacologically active in humans (see chap. 5). There are other opportunities later in the process to perform decarboxylation; however, this may require additional equipment or processing time.

Hammer mills are ubiquitous in cannabis and hops processing to grind biomass into a uniform size for downstream extraction. Any material that contains glandular trichomes can be subjected to extraction; the concentration of trichomes by weight is a primary determining factor in the yield of terpenophenolics and terpenoids. Extractors looking to create high-CBD concentrates from industrial hemp tend to use inflorescences, fan leaves, and stems to maximize CBD yields. Marijuana producers often use trimmed fan leaves and sugar leaves, plus smaller bits of inflorescences that are a waste product from harvesting. Visually appealing marijuana inflorescences can command high prices in regulated stores, less so with the smaller bits. After grinding, the biomass is packed into an extraction column in preparation for the extraction step.

Step 2: Solvent Extraction

Extraction concentrates the cannabinoids and terpenes from the biomass. The process utilizes a solvent that dissolves the cannabinoids and terpenes and the solute-laden solvent is then separated from the biomass by filtration. The solvent is then evaporated, leaving a highly concentrated extract of cannabinoids and terpenes.

Extractor operators tailor their apparatus to the unique properties of the solvent. Most extractors use high pressures to induce phase changes in the solvent, moving it between liquid and gaseous phases. This allows extractors to remove the solvent after extraction through evaporation and to recollect the solvent by re-liquifying it under pressure.

The two primary solvent types used in the cannabis industry are liquid carbon dioxide and hydrocarbons such as hexane or butane. Liquid carbon dioxide is becoming the preferred solvent in both hemp and marijuana industries because it is food grade, non-flammable, and does not leave residue in the final product. Butane extracts tend to have specific applications in the marijuana industry—in regulated and clandestine markets—for vaporizers and other specialty concentrates meant for smoking and vaporizing. Butane has been widely used because it is cheap, easily accessible, and requires less expensive equipment to get started, although it may be more expensive in the long run. Additionally, legacy knowledge from former black-market extractors who went "legitimate" in the early days of legal recreational marijuana were familiar with using butane, leading to its adoption.

The use of hydrocarbon-based solvents for extracting cannabinoids to be used in food products has been mostly eradicated in the regulated market but can sometimes still be found in the gray and black marijuana markets. Hydrocarbons, if ingested even in small quantities, can be neurotoxic and carcinogenic (Romano and Hazekamp 2013, 6), and low-grade supplies can carry impurities that are also toxic. Hydrocarbon-based solvents such as butane do not always completely volatilize during the extraction process and can leave residual solvent in the concentrate. In addition to the health risks, explosions from improper design and operation in butane extraction facilities have discouraged new extraction businesses from adopting the practice (Rainey 2019). Extractor operators in legal markets must install expensive fire and explosion suppression equipment to meet safety requirements, driving up expenses.

The combination of safety regulations and residual solvent testing requirements within legal, regulated marijuana markets pushed many extraction operations to switch to supercritical carbon dioxide to guarantee quality compliance. During the transition from the early days of the marijuana business to hemp legalization, operators working to extract industrial hemp-derived CBD mostly adopted food-grade extraction solvents from day one and avoided many of the problems of the early regulated marijuana concentrates. Liquid carbon dioxide extractors present their own safety challenges due to the extreme pressures and temperatures required to vaporize and liquify carbon dioxide.

Some CBD extractors use high-proof ethanol at atmospheric pressure. In this case, the process combines the solvent extraction and winterization steps; however in between the two steps, there is a filtration step to remove the non-extracted solids prior to winterization.

Step 3: Winterization of the Extract

After solvent extraction and subsequent removal of the solvent, the cannabinoid concentrate is now in a form called "crude oil," containing basically anything that solubilized. This includes not only the cannabinoids and essential oil—roughly 60–85 percent of the extract—but also, waxes, fats, chlorophyll, and lipids. The latter compounds can cause stability and taste problems, so extractors "winterize" the crude oil to refine a purer, more stable form. Essentially, winterization separates the extract components using cold temperatures, typically between −4°F and −112°F (−20°C and −80°C). To aid in this separation, the operator will first dilute the crude cannabinoid oil with a liquid solvent, typically ethanol because of its efficacy, safety, availability, and cost-effectiveness. The ethanol must be of sufficient proof, otherwise it will poorly solubilize the extract. Food-grade ethanol of 70 percent (140 proof) or greater is common, as are commercial products like Everclear.

Once the crude oil is dissolved in the solvent through mixing, it is then cooled. With the proper temperature and time, the waxes and fats form a separate layer, which can then be filtered. The final step is to remove the solvent, typically by distillation. Many cost-conscious extractors wish to recapture the ethanol for reuse in future batches, so the cannabinoid oil and ethanol mixture is transferred to a distillation apparatus for ethanol removal and capture. Additionally, the captured ethanol accumulates essential oils from the concentrate and extractors must take care to manage the purity of the recovered ethanol to not contaminate the organoleptic properties of future extract batches. This removal step is another obvious time at which to decarboxylate the concentrate; however, decarboxylation occurs at temperatures greater than ethanol vaporization so they would have to be two distinct steps.

You may be wondering why the operator couldn't just leave the solubilized concentrate in the ethanol and sell that product. Setting aside the legal challenges, there are also technical challenges in using an ethanol-solubilized concentrate in beverage formulation. If you were to dose this ethanol–cannabinoid concentrate into a beverage, the ethanol dilution would cause the hydrophobic cannabinoids to precipitate out of solution. Depending on the rate of dilution, the precipitation, or "outbreak," may be instantaneous or it may take some time. By the time the beverage was packaged, shipped, and made it home, the cannabinoids would likely fully separate out from the beverage. This separation would make the beverage look unappealing, plus the cannabinoids may cause flavor issues.

At this point, the extract is now in a finished and stable form and can be blended with terpenes or flavors and used to fill vaporizer cartridges, mixed into confections and other food products, or, for our application, mixed into a beverage through emulsification.

Step 4: Preparation of Extract and Emulsification

Formulating a cannabinoid beverage begins with taking the refined cannabinoid oil and binding it to a food-grade carrier oil. Concentrated cannabinoids can be tricky to handle, mostly because they are incredibly sticky and harden at low temperatures. Cannabinoid concentrates can be heated to reduce viscosity and get them to flow, but the more common method is to bind them to food-grade oils. Theoretically, it is possible to create a stable cannabinoid emulsion without a carrier oil, but in practice it is uncommon due to the challenges presented by the viscosity. The process of dissolving cannabinoids in oils is rudimentary and common in edibles—think marijuana cooked in butter or olive oil.

The cannabinoid concentrate will dissolve into the carrier oil; gentle heat speeds up this process. Typically, this is one of the first quality checks where cannabinoid content is measured, usually via high-performance liquid chromatography (HPLC). From this point forward the cannabinoid concentrate will be further diluted and incorporated into an emulsion, so this is the last step where we can assume we have a homogenous solution and is a logical place to establish a baseline for cannabinoid content to guarantee dosing consistency. After establishing the potency of the concentrate dissolved in carrier oil, a manufacturer will blend the cannabinoid-rich carrier oil with the primary emulsifier.

A stable emulsion becomes possible with the use of an emulsifier, which is a type of surfactant. Food scientists tend to use the terms *surfactant* and *emulsifier* interchangeably, but emulsifiers are one type of surfactant, one that encapsulates an immiscible substance, allowing it to behave in the bulk solvent as if it were soluble. Emulsifiers allow the cannabinoid-rich oil to be suspended in a beverage and protect the cannabinoids from oxidative degradation. As with all surfactants, the molecular structure of an emulsifier has two distinct components, a hydrophilic (water-loving) "head" and a hydrophobic (water-hating) "tail." The tail of the surfactant will extend into the cannabinoid oil droplet and the head will form a protective bead across a section of the oil droplet surface. With the right ratio of surfactant to oils, the hydrophilic heads will surround the oil droplet entirely, allowing it to stay suspended in the beverage. This ratio of surfactant to oil is represented by the hydrophilic-lipophilic balance (HLB), a measurable property that expresses a surfactant's binding affinity with different ratios of oil and water. Depending on the emulsifier, additional stabilizers may be added to the primary emulsifier. Stabilizers act as their name suggests: to make sure the primary emulsifier stays in a stable state. Instability of an emulsion can happen as a result of pH, metal ion interactions, oxidation, ultraviolet light, heat, or extended periods of time in storage.

We now have a coarsely emulsified cannabinoid oil that will suspend in a beverage; however, it will not be stable long term. The variability in the size and composition of the encapsulated oil droplets increases the odds of interaction between the particles or between the particles and the beverage, so they must be further processed into a uniform size. A useful analogy for brewers is producing a stable hazy beer. When a hazy beer is stable, it is a beautiful thing to behold, if not, it looks like the world's most unappealing snow globe. The same concept applies to cannabinoid emulsions. Instability in emulsions can take the form of globulation (oil droplets coalescing), flocculation (droplets settling), or separation (slicking). Even if we were to accept the visually unappealing components of an unstable emulsion, the encapsulation system also protects components of the cannabinoid oils from oxidizing and degrading, which can lead to flavor complications like rancidity and loss of valuable cannabinoid or flavorant content. Upon dosing this unstable emulsion into your packaging tank, you run the additional risk of package-to-package variability in terms of cannabinoid content, exposing you to further legal and financial liability.

Emulsion Types

Micelles. Micelles are thermodynamically stable molecular aggregates formed from surfactant molecules containing hydrophilic head groups and hydrophobic tail groups. In a water-based (aqueous) liquid, micelles form spontaneously when the hydrophobic tails collide and bind with one another, eventually self-organizing into a spherical orientation where the tail portions face inward, completely excluding water, and the heads face outward into the surrounding water-based liquid phase, completely shielding the core (see figure). Should a small hydrophobic particle, such as a cannabinoid-rich carrier oil droplet, come into contact with one of these agglomerations, it is thermodynamically favorable for the particle to be sequestered in the micelle core with the hydrophobic tails. Provided the surfactant (emulsifier) exists in sufficient concentration, micelles will spontaneously self-assemble due to the effect of hydrophobic interactions (McClements 2014, 16). Well-formed micelles tend to distribute well in water and have a high hydrophilic to lipophilic balance, due to the high degree of encapsulation by the hydrophilic heads around the hydrophobic core. When an oil droplet is incorporated into a micelle these are often termed "oil-swollen micelles" due to the diameter of the micelle increasing to accommodate the oil droplet. Swollen micelles are often interchangeably referred to as "emulsions" or "nanoemulsion droplets." Micelles are very small; they are usually processed into nano- or micro-scale particles—less or greater than 200 nm in diameter, respectively.

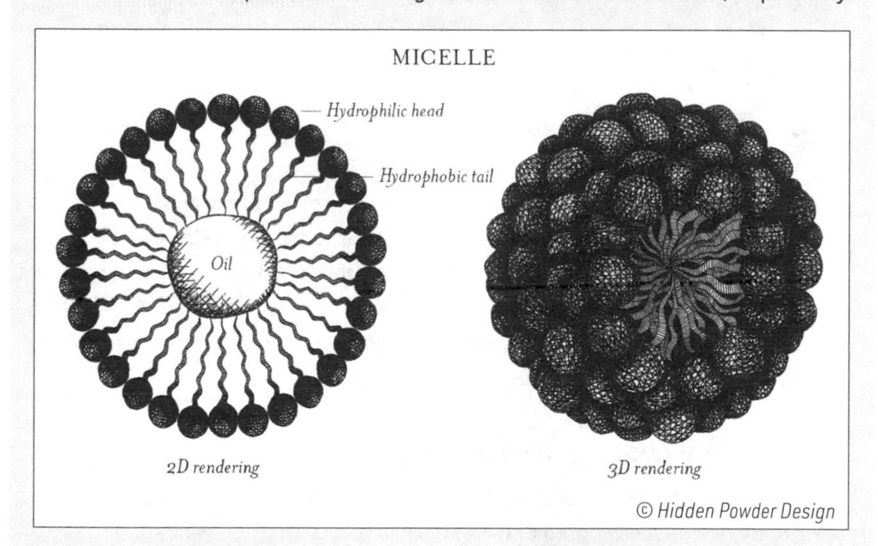

Liposomes. Liposomes are composed of bilayered surfactant molecules, where the hydrophobic tails of each layer face one another (see figure). Liposomes are spherical, but note that the center of a liposome is a pocket lined with the hydrophilic heads of the surfactant molecules; hence this enclosed pocket is a hydrophilic environment. Liposomes are formed from a unique surfactant class called phospholipids, which contain a single hydrophilic head and two hydrophobic tails. Liposomes can be developed to emulsify both hydrophilic and hydrophobic bioactive ingredients, making them particularly useful for the formulation of pharmaceuticals (McClements 2014, 17–18).

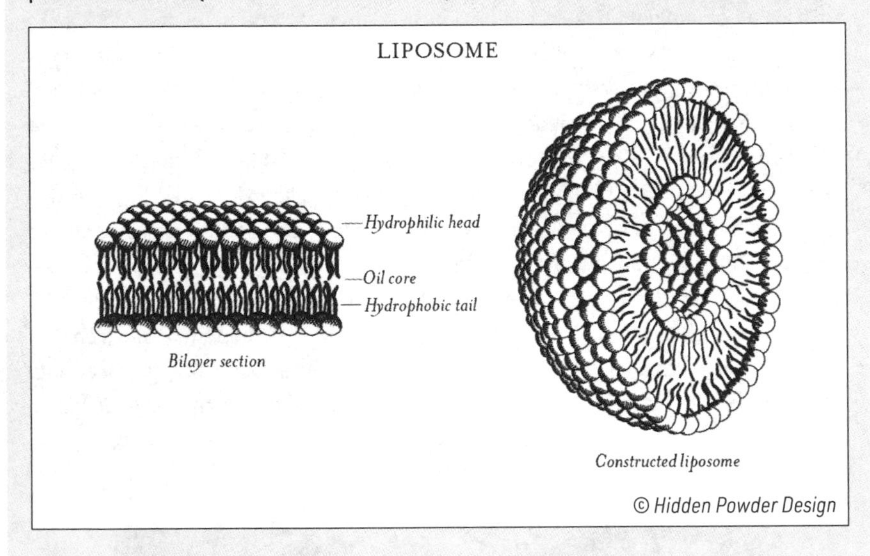

Solid Lipid Nanoparticles. Solid lipid nanoparticles look and behave similarly to micelles, except that the hydrophobic core is solid, rather than liquid. The pharmaceutical industry developed solid lipid nanoparticles to overcome some of the bioavailability issues associated with the other emulsion systems. In many applications, solid lipid nanoparticles can protect against digestive degradation and are suitable formulations for gradual release of bioactive compounds such as cannabinoids (Severino et al. 2012).

Common Emulsifiers (Surfactants)

The following common emulsifiers (surfactants) are generally recognized as safe (GRAS) for use as food ingredients in the US and EU. They are all intended to be oil-in-water carriers.

Gum Arabic. Gum arabic is a common food and beverage emulsifier made from the sap of one of several acacia species. It is a polysaccharidic emulsifier.

Soy Lecithin. Soy lecithin is a common emulsifier made from a mixture of phospholipids extracted from soybean oil. It is a lipid-based emulsifier.

Quillaja. Quillaja is an extract made from the inner bark of the South American soap bark tree. This extract is rich in saponins, which are triterpene glycosides than can act as powerful emulsifiers.

Polysorbates. Polysorbates are the esterified by-product of the sugar alcohol sorbitol, a common sugar substitute.

Step 5: Homogenization of the Emulsion

Emulsion stability depends on several factors. First, the suspended particles must be of uniform size and dispersed evenly throughout the continuous phase (the beverage). Second, the density and chemical composition of the emulsion must also be similar to the continuous phase. Finally, the continuous phase must be chemically stable throughout the shelf life of the beverage. Most emulsions will not naturally form into uniformly sized particles, so beverage manufacturers must use the process of mechanical homogenization to achieve stability.

Once the cannabinoid system is emulsified, it is ready to be homogenized. This is the step that produces a cannabinoid concentrate that is ready for packaging and shipping to your beverage-making facility. Homogenization is the process that reduces the emulsified particles to a smaller, uniform size using mechanical action and shear forces (see sidebar p. 138). This is a crucial step in achieving stability of the cannabinoid oil emulsion and sets up the beverage maker for success when they add it to their product. Without homogenization, you risk major emulsion instability driven by differing sizes of oil droplets and emulsified complexes. Differing sizes of active particles in a beverage tend to interact with one another, aggregating via various mechanisms to create even bigger particles, which can then form chunks and settle out.

Generally, the smaller the particle diameter in an emulsion, the more likely it is to be stable. Most cannabinoid emulsions exist either as nanoparticles (less than 100 nanometers) or as microparticles (less than 1,000 nanometers, i.e., less than a micrometer). These dimensions are chosen for a variety of reasons. While it is certainly possible to have an emulsion stable enough for use in a beverage with emulsified microparticles, emulsion stability greatly increases as you reduce the particle size down to nanoparticles. This is particularly important for beverages that will sit on store shelves for a long period of time or that may experience significant stress in the supply chain. Nanoparticles also tend to have greater bioavailability than microparticles and are sometimes chosen for that reason alone. Often, however, the choice between formulations is driven by clarity: for waters, teas, and clear beers, nanoparticles allow for visual clarity due their inability to effectively scatter light, whereas microparticles—in stable and unstable forms—throw a haze into a beverage.

Homogenization Techniques

High-Sheer Mixing. The most rudimentary homogenization technology is high-sheer mixing, which uses a high-speed impeller to blend something into a mostly uniform solution. Think of these devices as high-powered blenders. High-sheer mixers work at atmospheric pressure, can mix both in-line and batchwise, and typically have the lowest input costs of any homogenizer (Tomke and Rathod 2020, 42). High-sheer mixing will reduce oil droplets down to microparticles; however, it cannot make nanoparticles.

High-sheer mixing tends to be a bit crude; it generates tremendous sheer forces in the liquid to minimize emulsified droplet size, but it also generates heat. This can lead to shelf stability problems, especially for emulsions carrying terpenes and aromatics. Oftentimes, high-sheer mixers are used in a premix, where the emulsion is roughly homogenized and then reduced to its final particle size by a more precise homogenization technology.

High-Pressure Homogenization. A high-pressure homogenizer uses extremely high pressures (500–5,000 psi, or 3.45–34.5 MPa) to force an oily solution and an aqueous solution into suspension. The two solutions are pumped under high pressure through a small-orifice plate, creating high amounts of turbulence and velocity. As these two solutions impact one another, the sheer forces from the turbulence break apart the oil solution into progressively smaller droplets, forming a nanoparticle emulsion.

There are different designs of high-pressure homogenizers, most notably one called a microfluidizer. The microfluidizer utilizes an interaction chamber where oil and water collide through a torturous path at high pressure. The resulting cavitation reduces the suspended particles to the nanoparticle emulsion scale. Because of the forces during high-pressure homogenization, a large amount of heat is generated. This can cause instability in ingredients that do not have resistance to thermal stress.

Ultrasonic Homogenization. High-amplitude ultrasonic homogenizers create emulsions by using acoustic waves to generate intense shear forces. The generators use ultrasonic frequencies to direct waves into the fluid and cause cavitation. The resulting energy transfer creates intense sheer forces and reduces the suspended droplets into micro- and nanoparticle dimensions. Not only will ultrasonic homogenizers produce stable emulsions, the process will also kill microorganisms in solution, allowing for aseptic production.

Importantly, the homogenization process can be manipulated to meet the specifications of the final beverage it will be dosed into. Myriad variables determine best practices per any given beverage and emulsion system; the better you know the physiochemical properties of your beverage, the greater the likelihood you can control the stability and shelf life of your cannabinoid-containing beverage.

Step 6: Dosing into the Final Beverage

There are many additional considerations required to successfully introduce a stable, emulsified concentrate into a beverage. First, the manufacturer must provide a stable system that can deliver a consistent product following the rigors of packaging, shipping, and storage for a reasonable time. Then, when ready to use, the concentrate must form a stable emulsion in whatever it is dosed into. Additional stabilizers to inhibit instability are common in these types of emulsion systems to guarantee emulsion stability throughout the process.

Concentrates can be made in both powdered and liquid form, depending on the needs of the final beverage. Powdered concentrates can be either mixed in with other beverage powders for mixing at home or mixed into non-carbonated beverages in manufacturing. To achieve a powdered concentrate, the homogenized cannabinoid emulsion is sprayed onto a water-dispersible powder and tumble-dried to evenly mix it. The dry powder desiccates the cannabinoid oil and allows it to be relatively stable in dry, ambient storage conditions. To use a powdered concentrate, simply rehydrate the powder by adding it directly to the beverage or by rehydrating it in deaerated water and then dosing into the beverage; use a method that suits the size and scope of your facilities.

Liquid concentrates allow for easy dosing and rapid dispersion into the final beverage. Sometimes the concentrate manufacturer will add water and/or propylene glycol to help with rapid dispersal upon dosing into the beverage. Packaging and storage of liquid concentrates require extra attention. Typically, the concentrate is packaged and stored in an aseptic container flushed with nitrogen or carbon dioxide. To protect against oxidative degradation, it is best to store the liquid concentrate containers cold; however, they may be stored at ambient temperatures if the containers are properly sealed.

The final check is guaranteeing an even emulsion distribution. In the regulated marijuana market, beverage manufacturers are allowed small variations in cannabinoid concentration from bottle to bottle, otherwise the batch is out of compliance and must be destroyed. In Colorado, for example, the acceptable variance is 15 percent or less of the stated cannabinoid content. Ensuring even mixing of the batch so that each packaged unit contains an exact amount of cannabinoid is crucial to commercial manufacturing success.

FACTORS DRIVING INSTABILITY FROM PURCHASED CONCENTRATES

Now that we have a packaged cannabinoid concentrate, we can shift our attention to the factors that drive instability through the end beverage's shelf life. While many of these factors apply upstream of this point, we are going to assume that the concentrate manufacturer took all steps possible to eliminate these hazards from their process.

There are many factors to consider when thinking about emulsion instability as it relates to dosing, shipping, and storage of your final beverage. One of the most common factors are the degradation reactions that most brewers would associate with flavor instability in beer. Excessive heat, moisture, oxygen, light, changing pH, and the presence of metallic ions in the beverage can all catalyze the degradation of an otherwise stable emulsion. Many of these factors can be solved with adequate process control; for example, minimizing dissolved oxygen in your finished product, auditing your distribution supply chain to mitigate controllable environmental stressors, or considering opportunities to minimize the uptake of metal ions (e.g., iron) when using diatomaceous earth during filtration. While all these factors may play a role in destabilizing emulsions, they do not guarantee it will happen. Shelf-life testing under stressed environmental conditions will help determine the shelf stability of your cannabinoid emulsion and must be conducted before launching a new product. The consequences of instability may not be limited just to emulsion separation, but could also introduce detrimental flavors, change the color, or oxidatively degrade the bioactive ingredients.

Luckily, additive solutions—often called stabilizers—can protect your emulsified cannabinoids against oxidative degradation. You can specify with your manufacturer to incorporate these additives in your emulsion product, or add them yourself when dosing the concentrate into your beverage. Certain natural polymers minimize the surface interaction between the beverage and the encapsulated cannabinoids. Different protein, polysaccharide, and lipid-based emulsifiers can provide adequate protection against oxygen diffusion across the encapsulated membrane, provided it is stable and free of defects during manufacturing. Additional stabilizers can also add to this antioxidant

Types of Emulsion Instability

Flocculation. Flocculation is the process where fine droplets clump together (aggregate) and settle out into the bottom of a container.

Coalescence. Coalescence is the process where droplets merge together and form a larger single droplet. This droplet can either remain suspended or flocculate.

Ostwald Ripening. Ostwald ripening is the phenomenon where smaller droplets of hydrophobic oils disaggregate into the continuous phase and eventually deposit onto larger droplets, causing the droplets to grow. When the larger droplets grow to a certain size, it throws a significant haze. This can be demonstrated by the ouzo effect, where oil-soluble (hydrophobic) fractions of anise essential oils in ouzo disperse and then recoalesce into larger droplets when diluted with water.

Phase Separation. Phase separation is the process where an emulsion loses its miscibility and will separate fully from the solution, which also causes the loss of most of the antioxidant functions of the emulsion. This typically arises when the continuous phase is chemically unstable, causing changes in pH and activity. These changes react with the emulsifying agent and will break up the encapsulation. In the case of an oil-in-water emulsion, the oil will float on top of the water.

shell and so these types of product can be tailored based on the beverage's expected relative stress factors, allowing the beverage maker to use an appropriate additive formulation.

Assuming that a beverage emulsion has a relatively homogenous particle size, you can have a reasonably high degree of confidence it will be stable against coalescence and phase separation. For the next level of quality assurance, you can look at one of the most elegant ways to study emulsion stability: the zeta potential. To measure an emulsion's zeta potential, we must imagine we are looking at a single suspended droplet of a nanoparticle emulsion and how the surface of this droplet interacts with its surrounding environment on a molecular level. Variables such as extreme pH, the presence of metallic ions, and high levels of dissolved oxygen present opportunities for the beverage to interact and exchange ions with the surface of the nanoparticle emulsion. In a perfect world, there would be minimal ion exchange between the beverage and the nanoparticle emulsion droplets, but that is rare in practical applications. As a result, we can measure the zeta potential, which is roughly defined as the electrical potential of the emulsified droplets in suspension across the plane where they interact with their environment. A high zeta potential implies the droplet is surrounded by a shell of ions that are strongly attracted to the droplet as it is suspended in the continuous phase (i.e., the beverage). Because of this ion shell, any two droplets that come close to one another in the continuous phase will be repel each other due to the charge of the ion shells (electrostatic repulsion); thus, stability is maintained. A high degree of ion exchange between the nanoparticles in suspension and the beverage alters the zeta potential and potentially leads to a breakdown of stability. Ideally, each particle should be chemically stable enough that it will continue to repel other particles around it through electrostatic repulsion. If the particles can more easily chemically interact with one another, one of the types of emulsion instability arises.

Instability can be studied using a technique called electrophoretic light scattering, in which an apparatus applies a low-voltage electrical current to a sample and then measures the speed and the direction of suspended particles in a solution by the light the particles scatter and absorb. The machine uses a wide field of light spectrum reading, combined with advanced computational methods that correct for Brownian motion (the random motion of suspended particles colliding with one another) to ascertain the distribution of particle sizes in solution, along with their electrical potential. A great piece of technology to check out is Anton Paar's Litesizer, an apparatus that uses a variety of light scattering techniques to measure the properties of emulsion systems (https://www.anton-paar.com/us-en/products/details/litesizer). It's mostly used in the pharmaceutical world but has applications for beverage makers as well.

BIOAVAILABILITY AND PHARMACOKINETICS

It is important to touch on the topic of bioavailability and pharmacokinetics to better understand the reasons for building different emulsion systems and the ways that they produce variable effects when consuming cannabis in a beverage. **Pharmacokinetics** is the study of how the body absorbs, distributes, metabolizes, and excretes a drug, whereas pharmacodynamics is the study of the biochemical and physiological effects a drug has on the body. Our digestive system is quite harsh and can often chemically dissolve and inactivate many compounds, potentially eliminating or attenuating any kind of psychoactive or pharmacological effect.

Much of modern cannabis research focuses on the concept of **bioavailability**, simply defined as the proportion of active substance that survives metabolism and elicits a physiological response. Food technologists and pharmacologists wish to increase overall bioavailability in cannabis products, either in pursuit of creating a more consistent and reproducible high or to increase the therapeutic efficacy of cannabinoids such as CBD. For humans to feel the pharmacological effects of cannabinoids, they must be absorbed into the bloodstream, where the cannabinoids can then be carried to the brain and interact with the cannabinoid receptors of the endocannabinoid system. Ultimately, the liver will metabolize cannabinoids to make them more polar (have regions that carry an electrical charge) so they can be more easily excreted in the urine.

Numerous studies into delta-9 tetrahydrocannabinol (THC) show that its bioavailability dramatically differs depending on the route of ingestion. For example, a study showed that the bioavailability of THC from smoking can range from 2 to 56 percent depending on how the smoke is inhaled, how much THC was destroyed when burnt, and the method of smoking (Perez-Reyes 1990, 42). Another study showed that oral ingestion of THC—in the form of a cookie—provided only six percent bioavailability; curiously, the study noted that the clinical effects were similar to smoking and lasted longer (Ohlsson et al. 1980, 411). This is due to the metabolism of THC, which is turned into 11-hydroxy-THC by the liver. 11-Hydroxy-THC is more psychoactive than THC and shows up in higher concentrations in the bloodstream when marijuana is consumed orally versus being smoked or absorbed transdermally.

Emulsion systems play a dual role of chemically stabilizing cannabinoids for ingestion and protecting cannabinoids from digestive degradation in the stomach. Ideally, as the body digests a beverage, digestive enzymes cleave and digest both the emulsifier and the carrier oil, freeing the cannabinoids to be absorbed and transported throughout the body before being inactivated by liver metabolism. Thus, not only is the choice of carrier oil important for factors such as the intended shelf life and how the base beverage interacts with the emulsion, it is also crucial to the intended pharmacokinetics and pharmacodynamics when the emulsion system is ingested.

Particle size is also an important factor when it comes to bioavailability. Generally, the smaller the particle, the more bioavailable it is due to the increased surface-to-volume ratio in the digestive system.

Thanks to the myriad variables relating to how each emulsion system influences the pharmacokinetics and pharmacodynamics of a cannabinoid beverage, when a consumer drinks different brands of THC beverages—on separate occasions—there is a possibility that each beverage will deliver a different effect, despite each brand having the same concentration of active ingredient.

Common carrier oils include both natural and refined oils such as olive oil, coconut oil, MCT (medium-chain triglyceride) oils, and LCT (long-chain triglyceride) oils. Refined oils tend to have better shelf stability; however, natural oils perform well and are preferred in organic and health-conscious formulations. Since we have a good idea of how different food-grade oils are metabolized in the body, manufacturers can use that knowledge to manipulate the pharmacokinetics and pharmacodynamics of the beverage. Many manufacturers talk about the onset and offset timing of edibles, basically how long it will take you to feel the effects and then how long to feel "normal" again. With alcohol, we know roughly how long it takes for a beer to kick in—roughly 5 minutes—and that it takes about an hour for the body to metabolize the ethanol. Compared to alcohol, cannabis edibles and beverages are far more challenging because the body not only has to metabolize and excrete the cannabinoids, but also the oils and carriers it is bound to. By the nature of the oils and carriers, it may be metabolized at different rates, thus it will vary in the time and way it kicks in.

7

STRATEGIES FOR DOSING CANNABINOIDS IN BEER

You should have already read the chapter on the theory behind how cannabinoids are solubilized for dosing into beverages (chap. 6). Recall from that chapter that cannabinoids are hydrophobic (or lipophilic), making them particularly challenging to disperse evenly into a beer-based system. There will be some useful information for homebrewers in this chapter but, for the most part, we are going to look at strategies to solubilize cannabinoids in a commercial brewery setting.

I should also state unequivocally that you should do your own research on the legal status of adding cannabinoids to your beer, especially if you are a commercial brewer. Chapter 11 will provide some additional resources on navigating the Alcohol and Tobacco Tax and Trade Bureau approval process.

We must do several things to make cannabinoids available and bioactive. If you took raw hemp flower and simply added it as a dry hop addition it will smell nice, but little, if any, cannabinoids will solubilize. What little cannabinoids do solubilize will still be in their acidic state (see p. 114) and not bioactive in the body. Any strategy to effectively dose cannabinoids must use decarboxylated preparations, otherwise you are simply using cannabis as a flavor. There are quite a few articles in the homebrew press about how to brew with *Cannabis sativa*—mostly marijuana—for combining alcohol and THC. Many homebrewers report that simply decarboxylating the inflorescences in an oven and dosing those into beer, either as a kettle or dry hop addition, yields positive results. I would like to dissuade you from following that advice. First, while it is true that some cannabinoids will transfer into your beer, they will by no means be physically or chemically stable. Cannabinoids will naturally settle during fermentation or storage and may precipitate out with your process waste, including dry hops, yeast, trub, and cold break. Some cannabinoids may survive the brewing process and make it into the finished package but will yield unpredictable results. Stored beer—in bottles or kegs—will create further instability; in the instance of kegs, this may concentrate your cannabinoids, possibly giving you successive glasses with the first ones having a high concentration of cannabinoids, then ones with a very low concentration. Homebrewers may tolerate this level of variance, but I encourage you to exercise caution when serving these beers to yourself or your friends. Especially in the case of THC, you or your friends may end up with a very different experience than you intended.

If you are a commercial brewer or beverage manufacturer, it's a bit more straightforward: there are cannabinoid labeling requirements for each serving and you must be within the mandated range of variance to legally sell your beverage. Failing to do so will cause you to lose your customers and, if you miss your labeled dosage, you can lose your business as well. Beyond that, I would argue that one of the biggest factors that will drive mainstream acceptance of cannabis is the repeatability of experiences. With alcohol, the consumer has a relatively good sense of what any individual drink will do to them, whereas, with cannabis products, the same dosage may yield different experiences depending on its preparation, form of consumption, and quality control parameters. The quicker the cannabis industry in general can make experiences repeatable, the quicker cannabis can be accepted by reticent consumers.

We will focus on cannabidiol (CBD) as our main cannabinoid in this discussion but know that, from a biochemical and processing perspective, any cannabinoid will behave similarly to CBD. While it may seem a little counterintuitive, this chapter will also include strategies to *avoid* solubilizing cannabinoids, as there are additional regulatory compliance considerations that go along with combining an alcoholic beverage and cannabinoids.

During the early days of commercial CBD products, many companies did not bother to calculate how much CBD they were adding, did not homogenize their batches, or deliberately left out CBD from their formulations. In fact, the FDA tested many CBD products in the market and concluded that a significant number of products contained no CBD at all! Adding cannabinoids like CBD with precision will be critical for your product quality and will earn the trust and admiration of your customers.

DOSING FOR CBD USING HEMP INFLORESCENCES

Adding hemp inflorescences (hemp "flower") or hemp extract to the brew kettle will seldom lead to effective transfer of cannabinoids into wort. So why even bring it up? First, there is a considerable amount of discussion on homebrewer websites about adding cannabis to beer and I would like to try to provide some clarity and context to the information out there. A common refrain is that the brew kettle can be an effective location to both extract aroma-active terpenes and to decarboxylate cannabinoids from raw flower. While this seems intuitive, this advice will mostly result in ineffective transfer and solubility of cannabinoids. The decarboxylation rate of CBDA to CBD at boiling temperatures is low—decarboxylation is more effective at higher temperatures than boiling can achieve. Most CBD extractors decarboxylate between 250°F and 280°F (121–138°C). Home and commercial brewers certainly cannot boil that hot. If you were to attempt decarboxylation at boiling temperatures, you would need to maintain the boil for over five hours to effectively decarboxylate CBDA to CBD. Perhaps if you were making a deeply caramelized Belgian-style quadrupel this could be effective; however, most worts will suffer deleterious effects when boiled for that long. Regardless, adding CBD in the boil kettle will more than likely not be effective. A significant amount of the CBD will precipitate out of solution during beer processing, much in the same way that isomerized alpha acids also come out of solution in hoppy beers. There are opportunities to lose cannabinoids in the hot break in the kettle, in the cold break and yeast cake during fermentation, and again during racking and maturation of your beer. Some quantity may survive, but it is tough to tell how much and how reproducible the process will be.

To be fair, much of this online advice is based on trial and error from hobbyists brewing with marijuana at home. A recent paper studied the decarboxylation rates of THCA, CBDA, and CBGA and found that each cannabinoid has a different optimal decarboxylation temperature. The paper found that THCA decarboxylates quicker at lower temperatures than CBDA, so there is some evidence to support anecdotal homebrewer accounts. The authors showed it is possible to mostly decarboxylate THCA to THC at 203°F (95°C) over about an hour (Wang et al. 2016, 266). I would recommend reading the article if you are interested in researching more, it is published in an open-access journal.

In a homebrew setting, adding decarboxylated hemp flower for CBD during dry hopping may yield some cannabinoid in the finished beer but it may not be evenly dispersed. As a homebrewer, if you want to persist with this type of addition, focus on adding the decarboxylated flower as late in the process as possible. This will allow the ethanol present to partially solubilize some of the CBD and, with a bit of luck, it will not prematurely precipitate out of solution before packaging. Do not allow a contact time of more than 48 hours on the material, and immediately bottle the beer. This will give the greatest shot of extracting and keeping CBD in solution and, even if the CBD precipitates in the bottle, it can at least be consumed at the intended dosage.

As I have mentioned already in reference to a homebrew setting, when employing processes with low efficiencies in terms of cannabinoid transfer and—more importantly—that lead to inconsistent dosages, the end product should be treated with caution. For homebrewers, I would recommend dosing using other products that are easier to measure, such as commercially available CBD nanoparticle emulsions or home-produced tinctures, the latter of which will be covered shortly.

BATCH DOSING CBD

We will first discuss dosing CBD as it relates to commercial manufacturing operations, and the perspective will be for several different sizes and degrees of manufacturing sophistication. If you are reading this as a homebrewer, please feel free to skip this section and pick things up at the "Point of Dispense Dosing at Home" section (p. 153).

Before we start, references to "water-soluble CBD," "CBD concentrate," or just "CBD" all refer to emulsified CBD nanoparticles, which were discussed in chapter 6. If you have a product that is not produced in the manner described in chapter 6, the below information will not yield good results.

Adding water-soluble CBD in the cellar can be highly effective depending on the processing equipment (e.g., centrifuges and filters) used to prepare beer for packaging. Strategies for adding CBD should focus on properly preparing the water-soluble CBD for dosing into your beer and to effectively disperse/mix it evenly prior to packaging. Dosing and mixing strategies should consider three criteria: maintenance of stability of the CBD concentrate emulsion, minimization of oxygen pickup, and, especially, even dispersal of CBD throughout the tank.

Generally, you should look to add your CBD as late in the brewing process as possible, ideally in your packaging tank. If you try to add CBD upstream of that, you run the risk of either misdosing or wasting CBD. You are trying to be precise with the amount of CBD that ends up in each package, so do not try to add it in with additions such as fruit, dry hops, or spices. Additionally, make sure your beer is stable and that you do not have to drain yeast, trub, or hops after you dosed your CBD.

The first step in the process is to calculate the amount of CBD you are adding to your beer to make sure it is in compliance with legal labeling requirements. Nothing is worse than dumping a noncompliant batch of beer because you added too much or too little CBD. Double- and triple-check that you are adding what you think you are. Then have someone else check your math. While you are figuring out how much you are adding, you will need two critical pieces of information: the volume of beer in your packaging tank and your desired concentration of CBD in the finished package.

While CBD concentrates can come in liquid or powdered form, it is best to dose your CBD using a liquid concentrate. Powdered concentrates should be hydrated with de-aerated water and mixed before adding to beer. Depending on recommendations from the CBD concentrate manufacturer, you may also wish to lower the pH of your water to between 4.0 and 5.5 with a food-grade acid before adding the powdered CBD. Some emulsions contain stabilizers that buffer against pH shocks and others do not. Lowering the de-aerated water's pH may lower the risk of emulsion instability, especially when dosing into a relatively acidic product like beer. Look at your CBD concentrate's safety data sheet to see the pH of the CBD concentrate before you dose it. If it is relatively close to your finished beer's pH, target your de-aerated water's pH to the emulsions. If the CBD concentrate pH significantly differs from that of your beer, shoot to adjust the pH roughly halfway between the CBD concentrate pH and your finished beer pH. As an aside, if you notice the CBD concentrate pH varies significantly from your finished beer pH, you may want to consider using a different product, or work with the manufacturer to get you a product better suited to your beer. It is imperative that you conduct benchtop tests with any CBD product before you commit to dosing it in your tank. It will save you lots of time, money, and stress.

Once you have the correct volume of CBD prepared, you must now focus on lowering the concentration of dissolved oxygen in the prepared CBD concentrate before dosing it into your finished beer. The first reason should be obvious to any brewer: excess pickup of dissolved oxygen in your finished beer will lead to physical and flavor instability, shortening the shelf life of your beer. The second reason is specific to cannabis: excess dissolved oxygen can potentially drive instability in the emulsion, potentially leading to oxidative degradation of your chosen cannabinoids into artifacts that may have less value to your consumer. The easiest way to lower the dissolved oxygen of your CBD concentrate is to add it to any kind of keg and hook up a carbon dioxide gas line to a modified and sanitized coupler (tavern head). At New Belgium, we pull the check valves out of a standard Sankey "D" system coupler and hook up a gas quick connect to the spear side of the coupler. This allows us to reverse the flow of gas down the keg spear, bubble the gas through the liquid in the keg, and vent gas out of the headspace through (what is normally) the gas-side of the coupler. Any method should gently bubble carbon dioxide through the CBD concentrate for at least 15 minutes to sufficiently scrub any dissolved oxygen out of the concentrate.

Once your CBD concentrate is sufficiently purged of dissolved oxygen, it is now ready to add to your beer. At this point, a plan should be made for thoroughly mixing and homogenizing the CBD concentrate throughout the batch so that each unit volume contains the same concentration of cannabinoid. Common mixing methods include adding the CBD concentrate to an empty tank and then transferring beer into the tank onto the concentrate; adding the concentrate through a sample port, then rousing the tank with carbon dioxide to mix; or hooking up a pump from the tank bottom to a stand pipe or racking arm and circulating the contents of the tank for an optimum period of time. Depending on the size and geometry of your tank, at least one, if not all, of these methods can be effective. Tank residence time post-dosing may also play a factor, as well as how long it will take to empty your tank during packaging. You may want to think of this homogenization step in a similar way to homogenizing a tank with sugar and yeast for package conditioning—during packaging, you ideally want some form of rousing or agitation to make sure everything stays in suspension, unless you can package the tank very quickly.

The only package that deserves special consideration is kegged beer. Since kegs dosed with CBD may experience settling over time, it is important to make a plan on how to maintain your CBD concentrate in solution, for example, by periodic rousing. Assuming you have a high-quality water-soluble CBD emulsion, this settling should be quite slow and should not present too many problems.

But, if you are shipping over greater distances or plan to deplete inventory over several months, it would be best to store the kegs inverted; that way when you look to tap the keg, any settled emulsion should fall back into suspension. If the kegs have sat around upright for a few weeks, you may want to tip or rouse the keg before serving the beer to make sure your customers get as consistent an experience as possible. Individual unit packages (i.e., cans or bottles) present less of a problem, because they are easily roused when it comes time to open and drink from them and the consumer will drink the whole container.

CBD Concentration Formulas

Start with your target CBD amount per unit finished package size. For this example, we will take a standard US 12 fl. oz. can or bottle, which we will convert to metric (12 fl. oz. equals 355 mL). So, if your target CBD dose per bottle is 25 milligrams (mg), your target is 25 mg CBD/ 355 mL beer. For simplicity when dealing with bulk volumes on a brewery scale, calculate that concentration per liter (L), remembering that 1 L is 1,000 mL:

$$\frac{25\text{ mg}}{1} \times \frac{1}{355\text{ mL}} \times \frac{1{,}000\text{ mL}}{1\text{ L}} = \frac{25{,}000\text{ mg}}{355\text{ L}} = 70.42\text{ mg/L}$$

Now that you have a dosing rate of 70.42 mg CBD/ 1 L beer, take the volume of beer in liters in your bright beer tank (i.e., your batch size) and calculate the total quantity of CBD needed for your batch:

$$\text{total quantity CBD needed} = \text{batch size in liters} \times \text{target CBD concentration in milligrams per liter}$$

From the example, if I had a 10 hectoliter (1,000 L) batch in my bright tank and wanted each 355 mL (12 fl. oz.) bottle to contain 25 mg of CBD:

$$\begin{aligned}\text{total quantity CBD needed} &= \text{batch size in liters} \times \text{target CBD mg/L}\\ &= \frac{1{,}000\text{ L}}{1} \times \frac{70.42\text{ mg}}{1\text{ L}}\\ &= \frac{70{,}420\text{ mg}}{1} \times \frac{1\text{ g}}{1{,}000\text{ mg}}\\ &= 70.42\text{ g}\end{aligned}$$

I will need to add 70.42 grams (70,420 milligrams) to my 10 hectoliter tank.

How to Make De-aerated Water

If you've never made de-aerated water before, it is quite simple. If you need a large quantity, it is best to fill a bright beer tank or fermentation vessel with cold water that is free of chlorine. Hook up a carbon dioxide gas line to a sintered carbonation stone in the tank and bubble carbon dioxide through the water for upward of 24 hours. Leave the vent line open so the scrubbed oxygen has a place to vent. Check the water with a dissolved oxygen meter until the water registers less than 10 parts per billion of dissolved oxygen.

IN-LINE DOSING OF CBD

In-line dosing of CBD is only feasible at facilities with sophisticated technologies to dose exact quantities of a given additive into transferring beer. Dosing can occur at any point late in the process, such as after a centrifuge polish, during filtration and carbonation, or even in-line from a bright beer tank to a packaging line. For still beverages such as waters, wine, mead, and spirits, it is possible to dose a water-soluble CBD concentrate directly into a package; however, it is inadvisable to do so for carbonated beverages due to potential foaming issues during packaging. For carbonated beverages, it is best to add your concentrate while the beverage is still uncarbonated or while the beverage is under enough pressure where gas will not break out of solution when dosing additives. In-line dosing makes the most sense for processing very large batches, or for facilities that have sophisticated technologies such as flash or tunnel pasteurization. With very large batches, it is challenging to maintain homogeneity throughout a very large tank, so it is essential to dose your CBD concentrate incrementally during a transfer. This guarantees that each increment of beer receives the same amount of concentrate without needing to constantly rouse or mix a large tank.

It is essential to coordinate your efforts with your CBD concentrate vendor to ascertain whether the emulsion has physical characteristics that require additional processing considerations, especially if the emulsion is unstable after pasteurization. Many companies can produce water-soluble CBD emulsions in an aseptic form, which can allow for brewers to sanitarily dose their concentrates post-pasteurization with an appropriate dosing system. Some emulsions do have thermal stability and can be pasteurized. Check with your vendor and make a testing plan before attempting to dose a large batch.

POINT OF DISPENSE DOSING AT HOME

If you are a homebrewer or just a tinkerer, you are probably reading the previous sections and thinking that none of this applies to you. Never fear, I haven't forgotten about you. Understanding the innate challenges of correctly solubilizing cannabinoids and reproducibly dosing them can also be tackled at home. Many of the products available to commercial brewers, such as water-soluble single-serve or concentrates, are also available for retail purchase. If using commercial products, most of the math is already done for you—just add however much you need, stir, and enjoy. This may be fine for folks that don't want a lot of fuss, but there's also a fun and relatively easy way to make an infusible cannabinoid tincture at home with very little specialized equipment.

Making a Tincture: Getting Started

Tinctures provide a homebrewer with endless options for customization: you can adjust the strength and potency as well as choose from a myriad of flavors to pair with all sorts of beers and beverages. To make a basic tincture, you will need:

- trimmed cannabis inflorescences
- 750 mL of a high-proof neutral spirit (e.g., Everclear or Bacardi 151)
- three 500 mL (pint size) Mason jars with lids
- coffee filter papers and a funnel
- measuring vessel that measures up to 500 mL
- scale

Before you begin, it is important to understand how potent your cannabis material is. This will allow you to have a precise understanding of the concentration of cannabinoids in your final tincture and will allow you to better adjust your dosage as you see fit. If you are purchasing commercial flower (i.e., inflorescences), the packaging on your flower should indicate the potency by weight, given as a percentage. If you grew the flower at home, do your best to estimate or research online the potency of your strain. Based on the potency, calculate the total potential weight of your cannabinoids by multiplying the total weight of your cannabis flower by the percent cannabinoids by weight. I will go over this calculation in more detail in the section below (p. 155), but just know that you will need this value to reliably and safely dose your beer using your homemade tincture.

Making a Tincture: Step by Step

Step 1: You will first want to decarboxylate your trimmed cannabis flower in the oven. Preheat your oven to 225°F (107°C). Alternatively, you can vacuum seal your flower and cook it in a pressure cooker if you are concerned about the smell.

Step 2: Grind or break up your cannabis flower into small bits and place on a foil-lined baking sheet.

Step 3: Place the baking sheet in the preheated oven and heat for 45–60 minutes. Remove the baking sheet from the oven and allow to cool to room temperature.

Step 4: Measure out your high-proof spirit. Measure and record the volume, as this will be needed (along with your total calculated cannabinoids) to determine your cannabinoid content per unit volume.

- *Note:* It is recommended that you use at least a 10:1 ratio of alcohol to flower to achieve proper extraction. Additionally, you do not want to use too much alcohol, otherwise you will over-dilute your tincture, causing you to drink quite a bit of alcohol when taking your dosed beverage with tincture added.

Step 5: Combine your decarboxylated cannabis flower and your measured high-proof spirit in a Mason jar. Seal the jar with a lid and store in a cool, dark place. Oxygen, heat, and light will all degrade cannabinoids over time, so consider these factors when storing your tincture.

Step 6: Every hour or two, shake the jar to speed up the extraction process. Extract for at least four to six hours, and you can go up to a week. The longer you extract, the more herbaceous plant flavor you will extract along with your cannabinoids, but the extract will be more potent in terms of cannabinoid content.

Step 7: After extraction is complete, filter out the solid plant material through a coffee filter paper-lined funnel into a clean Mason jar.

Step 8: If you are consuming your tincture immediately, you may enjoy it now. If you plan on storing your tincture for an extended period, proceed to step 9.

Step 9: Place your tincture into the freezer to winterize and refine the extract. Let the tincture sit in the freezer undisturbed for at least 48 hours. You should notice a layer of waxy solids forming in your tincture either on the surface or suspended in solution. This is expected.

Step 10: Place another clean Mason jar, along with the funnel and coffee filter paper, into the freezer for at least two hours. The waxes and lipids that have separated are temperature sensitive, so you must keep all your equipment as cold as possible during the filtration step. You may need to make room in your freezer to do so.

Step 11: Ideally in the freezer, pour your winterized tincture through the coffee filter paper-lined funnel into the cold, clean Mason jar. You may need to work slowly in batches as the waxes can quickly clog the coffee filter paper. Keep extra coffee filter papers in the freezer in case they clog.

Step 12: Once fully filtered, store the tincture in the freezer, or at least in a cool, dark place.

- *Note:* Play around with different base products for producing your tincture. Using whole flower will lend some vegetal characteristics to your tincture, which in certain applications may be a welcome flavor addition. Concentrated trichome products like kief, hash, live rosin, and extracted concentrates will result in a less vegetal flavor profile and will extract a higher concentration of aroma active terpenes per unit weight relative to whole flower.

Calculating the Cannabinoid Content of Your Tincture

The decarboxylation reaction does not completely convert the acidic cannabinoids to neutral (bioactive) cannabinoids. Variables such as temperature fluctuations, thermal degradation of neutral cannabinoids during the decarboxylation process, and processing losses all should be accounted for when determining the total cannabinoid content of your tincture. A good assumption is that you will convert 80 percent of your acidic cannabinoids to their neutral state, but know that you may be off by roughly 10 percent on either side. As you will see, it should not cause too much variance in the final product. Ideally, have a cannabinoid analysis done on your raw flower so that you can calculate your tincture with greater accuracy; if that is not feasible, then research expected cannabinoid concentrations for your strain online. As with all the calculations so far in this chapter, I recommend performing your measurements and calculations in metric, as there will be less converting of units in the end.

Total cannabinoid content. Perform this step *before* you treat your cannabis material in the oven. Start by weighing out your raw cannabis flower on a scale and record the total weight in grams. Then multiply this recorded weight by the percent cannabinoids by weight of your flower, which should be provided in commercial samples (for homegrown, you will have to research the strain and do your best to obtain a reliable percentage). This will give you the total amount in grams of cannabinoid in your sample. Next, multiply the total grams of cannabinoids in your sample by 1,000 to convert the total into milligrams. This will help you later in calculating the potency of your tincture.

For example, 10 grams (g) of hemp flower that is 12 percent CBD by weight will contain 1.2 g of CBD, or 1,200 milligrams (mg).

$$\begin{aligned}\text{total cannabinoid content (mg)} &= \text{total weight cannabis (g)}\\ &\quad \times \text{percent cannabinoid by weight (\%)}\\ &\quad \times \text{conversion factor (mg/g)}\\ &= \frac{10\text{ g}}{1} \times \frac{12}{100} \times \frac{1{,}000\text{ mg}}{1\text{ g}}\\ &= 10 \times 0.12 \times 1{,}000\text{ mg}\\ &= 1{,}200\text{ mg}\end{aligned}$$

Decarboxylation loss. Let's assume, as mentioned earlier, that 80 percent of the acidic cannabinoid content is converted to neutral cannabinoid. Take your total milligrams of cannabinoids and multiply it by 0.8 (i.e., 80%, or 80/100) to give you your total expected yield of neutral cannabinoids available post-decarboxylation.

$$\begin{aligned}&\text{total expected yield neutral cannabinoids (mg)}\\ &\quad = \text{total cannabinoid content (mg)} \times \text{percent efficiency decarboxylation}\\ &\quad = 1{,}200\text{ mg} \times 80\%\\ &\quad = 1{,}200\text{ mg} \times 0.8\\ &\quad = 960\text{ mg}\end{aligned}$$

Concentration of tincture. Take your total expected yield of neutral cannabinoids and divide it by your measured volume of neutral grain spirit. This will give you your expected final concentration of cannabinoids, that is, the amount of cannabinoids in milligrams per unit volume of tincture, in this case milliliters (i.e., mg/mL). Let's assume, as mentioned in step 4 in the previous section, that you use a 10:1 ratio of alcohol to flower, so for 10 g of flower you use 100 mL neutral grain spirit.

$$\begin{aligned}&\text{cannabinoid conc. of tincture (mg/mL)}\\ &\quad = \frac{\text{total expected yield neutral cannabinoids (mg)}}{\text{measured volume neutral grain spirit (mL)}}\\ &\quad = \frac{960\text{ mg}}{100\text{ mL}}\\ &\quad = 9.6\text{ mg/mL}\end{aligned}$$

Worked example. Let's work through another example from start to finish. In this example, we will calculate the final concentration based on two different ratios of alcohol to flower; you may want to use more than a 10:1 ratio to ensure complete extraction of available cannabinoids. The cannabis material is hemp flower, so it is high in CBDA and low in THCA.

Starting material
10 g hemp flower
13% CBDA by weight
0.28% THCA by weight

Cannabinoid content:

$$\begin{aligned} \text{CBDA content} &= \frac{10\text{ g}}{1} \times \frac{13}{100} \times \frac{1{,}000\text{ mg}}{1\text{ g}} \\ &= 10 \times 0.13 \times 1{,}000\text{ mg} \\ &= 13{,}000\text{ mg CBDA} \end{aligned}$$

$$\begin{aligned} \text{THCA content} &= \frac{10\text{ g}}{1} \times \frac{0.28}{100} \times \frac{1{,}000\text{ mg}}{1\text{ g}} \\ &= 10 \times 0.0028 \times 1{,}000\text{ mg} \\ &= 28\text{ mg THCA} \end{aligned}$$

Calculating decarboxylation loss:
Assume 80% of the CBDA and THCA are converted to their neutral forms.

$$\begin{aligned} \text{yield of CBD} &= 1{,}300\text{ mg CBDA} \times 80\% \\ &= 1{,}300\text{ mg} \times 0.8 \\ &= 1{,}040\text{ mg CBD} \end{aligned}$$

$$\begin{aligned} \text{yield of THC} &= 28\text{ mg THCA} \times 80\% \\ &= 28\text{ mg} \times 0.8 \\ &= 22.4\text{ mg THC} \end{aligned}$$

Calculating concentration in final tincture:
First, we will stick to the minimum 10:1 alcohol to flower ratio recommended to achieve proper extraction. That's 100 mL neutral grain spirit to 10 g hemp flower.

$$\text{CBD conc. in final tincture} = \frac{1{,}040 \text{ mg}}{100 \text{ mL}} = 10.4 \text{ mg/mL}$$

$$\text{THC conc. in final tincture} = \frac{22.4 \text{ mg}}{100 \text{ mL}} = 0.2 \text{ mg/mL}$$

Second, let's suppose we increase the alcohol to flower ratio to 20:1. That's 200 mL neutral grain spirit to 10 g hemp flower.

$$\text{CBD conc. in final tincture} = \frac{1{,}040 \text{ mg}}{200 \text{ mL}} = 5.2 \text{ mg/mL}$$

$$\text{THC conc. in final tincture} = \frac{22.4 \text{ mg}}{200 \text{ mL}} = 0.1 \text{ mg/mL}$$

Of course, you will more than likely lose a bit of tincture during the extraction and filtration steps, so your total yield by volume will more than likely not be the same as what you put in. The calculated concentration should, however, stay the same and therefore you should not lose potency.

Flavoring Your Tincture

Depending on the *Cannabis sativa* strain or cultivar you used, you may get a wide variety of flavors that carry over into your tincture. Raw cannabis flavors in a tincture may complement certain types of beverages, but may detract from others. This presents a fun opportunity to manipulate the flavor of your tincture to better complement your beverages. Tinctures can be treated similarly to bitters in cocktail mixology: a few dashes can deliver a potent shot of aromatics along with the cannabinoids. These aromatics, in combination with the natural terpenes from the cannabis, can heighten certain aromas and provide a great pairing to the beer in which the tincture will be dosed. Since tinctures have a high ethanol content, it will extract herb and spice aromas readily and can be tuned to any flavor palette. Be mindful of both the ethanol concentration of the tincture and the concentration of cannabinoids within the tincture so you don't inadvertently overdose your drink with either ethanol or cannabinoids. As a final note, do not be surprised when you dose your tincture into your beer or beverage that it will turn cloudy or even milky. This is a demonstration of the ouzo effect, which involves Ostwald ripening (see p. 139). This is a normal and natural phenomenon when diluting a high-alcohol liquid containing oils down to a lower strength; the hydrophobic cannabinoids will coalesce into larger droplets, creating a haze in the beverage.

RECIPES FOR TINCTURES

ADJUNCTS ON ADJUNCTS: BARREL-AGED STOUT TINCTURE

This tincture will go beautifully with any dark beer, especially barrel-aged stouts. Adjunct barrel-aged stouts are all the rage these days; with this tincture, you can control which flavors you want in your stout.

INGREDIENTS

100 mL CBD tincture
2 large vanilla beans, split lengthwise with a knife
5 g high-quality cacao nibs
1 half-inch cinnamon stick

INSTRUCTIONS

Soak the vanilla beans, cacao nibs, and cinnamon stick in your CBD tincture for at least two days, possibly up to two weeks. Periodically shake the tincture to speed extraction. The tincture will be ready when it is strongly aromatic. You can adjust the spices to your preference. This tincture should keep for up to a year in the freezer.

IPAS FOR DAY-ZE: IPA TINCTURE

This tincture utilizes a unique preparation of citrus peel that will complement the terpenes in both cannabis and in hops. This recipe uses oleosaccharum (Latin for "sugar oil"), which helps draw out the essential oils from the citrus peel. It gives a very fresh and bright aroma of citrus and should go great with both the tincture and with any highly aromatic IPA.

INGREDIENTS

500 mL CBD tincture
Citrus blend oleosaccharum:
1 large grapefruit, peeled
3 limes, peeled
2 lemons, peeled
200 g (1 cup) cane sugar
60 mL (¼ cup) boiling water

INSTRUCTIONS

1. Peel the citrus fruits and place in a bowl. Try to leave the white pith on the fruit.
2. Add the sugar to the citrus peels and mix thoroughly. Let stand at least 6 hours. Stir once or twice to dissolve the sugar.
3. Strain the citrus peels into a clean container. Save the now drier peels, there's still good stuff in there!
4. After straining the primary liquid from the peels, add the boiling water to extract any residual sugar and citrus oils.
5. Strain the water into the same container as the rest of the oleosaccharum.
6. Blend the oleosaccharum into the CBD tincture and store in the refrigerator or freezer. For best flavor, use within two to four weeks.

WEEKDAY DE-STRESS: LOW-ABV COCKTAIL TINCTURE

This tincture is designed to be added to sparkling water or tonic water as a low-alcohol cocktail. The only alcohol will come from the tincture itself, so aim to add at most 5–10 mL of tincture. Even at the very high proof, the relative dilution of your tincture will yield a low-alcohol cocktail that's perfect for relaxing midweek but not leaving you hungover. The spices come from gin recipes; play around with it to get a flavor that is to your liking.

INGREDIENTS

500 mL CBD tincture
10 g juniper berries
peels of 2 lemons
2 g coriander seeds
1 g star anise (should be one whole one)
1 sprig fresh rosemary

INSTRUCTIONS

1. Soak all botanicals and lemon peels in your tincture stored in the freezer.
2. Shake the jar once a day and let extract for upward of two weeks, or until the tincture is sufficiently aromatic. Remove the botanicals and lemon peels.
3. Blend 5–10 mL of tincture into 355 mL (12 fl. oz.) of sparkling or tonic water, then garnish with a wedge of lime and serve.
4. This will yield a cocktail that contains between 1.6 and 3.2 percent ABV.

8

BIOCHEMISTRY OF CANNABIS OIL

Up to this point, we have been looking at cannabinoid production in the cannabis plant (chap. 5) and how cannabinoids can be infused into beverages (chap. 6), including beer (chap. 7). However, cannabinoids are not the only compounds secreted by cannabis glandular trichomes. Understanding the biochemistry of cannabis oil is essential for understanding how infusing cannabis compounds into your beverages will affect the final product. Many different types of aroma compounds, including esters, aldehydes, ketones, thiols, alcohols, and terpenes, comprise the oil fraction of cannabis glandular trichomes. If you have been to a marijuana dispensary, you've likely heard the term **terpenes**. We first encountered terpenes in chapter 4 (p. 89) as a major component of plant essential oils. Terpenes are unsaturated hydrocarbons commonly found in a variety of plants, including hops, and will be the main focus of this discussion, although we will return to some of the other essential oil constituents such as ketones, esters, and thiols.

Most brewers understand terpenes through their use of hops in brewing. Through this lens, this chapter will look at the subject of terpenes and their underlying biochemical characteristics. Chapter 10 will take this background knowledge and apply it to how a brewer would go about manipulating the terpenes found in cannabis to make delicious and unique beers.

Aside from cannabinoids, terpenes are one of the best studied subjects in cannabis biochemistry. Terpenes and their derivatives are the main chemical constituents that give cannabis its characteristic aroma. Terpenes are a naturally occurring and bewilderingly diverse class of hydrocarbons that exhibit a broad diversity of aromas. They are naturally produced by plants, animals, and microorganisms to serve a variety of biochemical purposes. Some terpenes do not accumulate and serve primarily as intermediary products, being used as building blocks for different molecules that an organism synthesizes. As discussed in the cannabinoid biochemistry chapter, the cannabis plant produces terpenes as one of the intermediary compounds in the synthesis of cannabigerolic acid (CBGA), the "mother" cannabinoid (p. 113). Being one of the largest classes of natural compounds, terpenes and their oxygenated derivatives can be found in everything from natural flavors, fragrances, cosmetics, perfumes, food additives, and pharmaceuticals.

Terpenes are made up of five-carbon units called **isoprenes** (C_5H_8) that are joined together to form terpenes of varying lengths $(C_5H_8)_n$, where *n* is 2 or more. The smallest terpenes are monoterpenes, formed by the joining of two isoprene units to make a ten-carbon (C_{10}-) molecule. Larger terpenes include sesquiterpenes (C_{15}-), diterpenes (C_{20}-), triterpenes (C_{30}-), and so forth. Generally, monoterpenes, along with some sesqui- and diterpenes, are highly volatile and have strong odors. As hydrocarbons, terpenes are naturally hydrophobic ("water-hating") and therefore lipophilic ("oil-loving"). As such, it is important to understand the properties of volatility and hydrophobicity when harvesting, processing, and drying cannabis ultimately for use in beverage applications.

Along with the many forms of terpenes, there are other terpene derivative products, broadly classified under the term **terpenoids**. Terpenoids, which possess one or more chemical groups containing oxygen, are typically oxidative by-products of terpenes and may have other functional groups stitched on to them. Terpenoids are highly prone to photo-oxidation due to their unsaturated nature. (Unsaturated means one or more carbons in the molecule have formed two bonds with the same atom—a "double bond"—usually another carbon or oxygen.) Terpenoids will readily transform into other oxygenated by-products such as alcohols, ketones, and aldehydes (Shapira et al. 2019, 11431).

OVERVIEW OF TERPENE BIOSYNTHESIS

There are two biosynthetic pathways that create terpenes in cannabis. The first is the mevalonic acid (MVA) pathway, which leads to sesquiterpenes and triterpenes. The second is the methylerythritol (MEP) pathway, which we were introduced to in chapter 5 as it is responsible for the biosynthesis of the monoterpene part of the "mother" cannabinoid CBGA (Andre, Hausman, and Guerriero 2016, 3). Both the MVA and MEP pathways are in service of producing isoprenoid precursors. However, the MVA pathway takes place in the cytosol (the organized fluid matrix that fills much of the cell) while the MEP pathway takes place in the plastids (a specialized organelle of plant cells); thus, the resulting isoprenoid precursors are exposed to different enzymes and are enzymatically reduced into different classes of terpenes. The MVA pathway leads to the formation of sesqui- and triterpenes and the MEP pathway leads to mono-, di-, and tetraterpenes.

The overall composition of the terpenes biosynthesized by cannabis is determined by genetics that code for different terpene synthase enzymes (Booth, Page and Bohlmann 2017, 2). Thus, most breeders and geneticists spend a lot of time looking for unique ways to breed for new and novel combinations of terpene synthases when looking to develop unique terpenes. Genetics informs how an individual strain of *Cannabis sativa* will biosynthesize terpenes that contribute to odor and flavor, but it does not determine the final composition of the essential oils. Environmental stress factors such as an abundance or lack of nutrients, predation, infection, weather variability, and drought all influence how terpenes will be expressed in the mature plant (Booth and Bohlmann 2019, 68).

There have been about 140 terpenes/terpenoids identified in *C. sativa* (Brenneisen 2007, 28), with more being discovered every year. Many terpenes in cannabis exist in very small concentrations in the plant and, for some terpenes, there may not be a reproducible standard for a laboratory to measure against. Additionally, with variable storage and handling procedures between producers, it can often be hard to ascertain whether the presence of particular terpenes or terpenoids can be attributed to the strain itself or if it is a physiological response to harvest, storage, and handling. We as brewers have the benefit of having a highly centralized, mostly uniform procedure for the harvest, drying, and storage of hops, leading to a high degree of reproducibility in hop terpene profiles. This is not the case in the cannabis business, where even a single strain can exhibit high variability due to lack of uniformity in growing conditions, drying times and practices, storage, and shipping. Small-scale efforts to limit oxygen and store material away from ultraviolet light sources have seen major improvements to quality in the cannabis market, but it still does not have the same rigor and uniformity that the hops market enjoys. This should be considered by any brewer looking to use cannabis as a flavoring agent in beer.

Using *Humulus lupulus* as our basis of comparison, *Cannabis sativa* produces similar levels of essential oil and terpenes on a dry weight basis to hops. It's more challenging to directly compare cannabis inflorescences to hop inflorescences because cannabis has secretory glandular trichomes on most of the plant, whereas the hop generally only has them on the inflorescences. Additional studies show variability of terpene production between fiber and seed varieties of industrial hemp chemovars versus marijuana chemovars.

Understanding this, various studies seeking to study the rate of essential oil production in *C. sativa*, both industrial hemp varieties and marijuana varieties, have reported essential oil contents ranging from 0.1 to 4 percent by weight. Confounding factors, such as seed set (less essential oil) versus unpollinated (more essential oil), fiber versus cannabinoid chemovars, and the raw material makeup of the sample (e.g., trimmed flower, leafy flower, or whole plant) help explain the large variability (Booth and Bohlmann 2019, 68–70; Mediavilla and Steinemann 1997, 81; Meier and Mediavilla 1998, 19; Ross and ElSohly 1996, 51). Concentrating the constituents of glandular trichomes, either through solvent extraction processes or more manual processes like creating hashish, will generally increase the essential oil content of the sample. One study showed that within the glandular trichomes themselves, essential oils make up to 10 percent of the total weight (Potter 2009, 161).

One study into hemp cultivars found that monoterpenes make up between 48 to 92 percent of the essential oil component, with myrcene and limonene making up the highest total concentration of individual terpenes (Mediavilla and Steinemann 1997, 82). Terpenoid (i.e., terpenes modified by being oxygenated) concentrations in cannabis inflorescences make up between 1 and 3.5 percent of the total weight of cannabis essential oils, a level comparable to hop cones (Fischedick et al. 2010, 2058).

TERPENE PRODUCTION DURING GROWTH, HARVEST, AND DRYING

With some basic knowledge of how terpenes are made in the plant, we can move on to understanding how the variables around growth and harvest timing can be manipulated to better control how the cannabis will smell and taste when it is harvested and dried. Through many conversations with the growers New Belgium worked with when developing the Hemperor brand, we often got the advice that we should look to harvest our hemp early because the farmers knew the smell was much more potent right before the cannabinoids started setting in.

Intuitively, this makes a lot of sense. Since terpenes are important biochemical building blocks, their levels will spike as the female inflorescences start preparing to reproduce. The essential oils of the cannabis plant play a protective role that is important to safeguard the stigmas and perigonal bracts of the inflorescence while it awaits the male pollen floating in the wind (discussed in chap. 4). As the days pass and the pollen doesn't come, the female plant will produce more flowers to up the odds of catching pollen. This in turn means more glandular trichomes secreting terpenes to protect the new flowers. Should the female inflorescences receive pollen, the production of terpenes tapers off and the plant now puts more of its energy and resources into growing viable seeds and preparing for winter dormancy or, more often, its death (Meier and Mediavilla 1998, 17).

Should no pollen come, in the waning days of summer the biochemical factories of cannabis kick into high gear. First, the plant will produce those primary terpenes and terpenoids that it has the specific terpene synthase enzymes for, which is dependent upon the particular variety's genetics. However, these terpene synthase enzymes may be expressed differently should certain environmental factors occur, such as a lack of rainfall (Booth, Page, and Bohlmann 2017, 3). These terpenes provide protective elements for the plant: repelling predators; inhibiting molds, bacteria, and fungi; and attracting pollinators. As the cannabis plant continues to produce its primary terpenes and terpenoids, exposure to oxidation, heat, and ultraviolet light will cause certain reactions to take place, producing oxidation events and rearrangements in the terpenes that lead to additional complexity in the plant's terpene profile. For example, the sesquiterpene β-caryophyllene oxidizes to caryophyllene oxide. Further transformation takes place during plant harvesting and drying. Drying the mature inflorescence, either through forced-air drying or by static-hang drying contributes additional thermal and oxidative stress. Many terpenes are highly volatile, so even just being exposed to open air or a large volume of headspace can diminish the concentration of certain terpenes significantly.

All the above is not inherently a bad thing, merely an explanation of the complexity of the enzymatic and non-enzymatic processes that happen when cannabis grows and matures. This should be quite familiar to commercial brewers, as we spend a considerable amount of time and energy studying, measuring, and selecting hops every year to meet our needs. The pitfalls of poor harvest timing and the sensory and organoleptic repercussions thereof provide a useful context to understand how we should go about managing cannabis production for future application in our beverages. Additionally, those interested in getting reproducible results with such applications should look to hop practices as a useful guide on how to get repeatable results at harvest. At present, these variables are poorly studied in the cannabis industry and significant research will have to be done to close this knowledge gap. Luckily, there are many other comparable use cases in a wide variety of industries that harvest plants to market the terpenes.

HOW TO MEASURE TERPENES

It is worth having a brief discussion on how to detect and quantify terpenes. Say you have a perfect sample of cannabis that smells amazing, and you can't wait to put it in your beer. How would you go about understanding that beautiful smell? Additionally, how could you confirm that the great smell of your sample stays that way until you are ready to brew with it?

The best way to measure terpenes is through a technique called gas chromatography (GC). Gas chromatography enables the measurement of volatile compounds by separating a sample into its individual components. Think of it as a very fancy distillation column that can measure minute changes in temperature and allows you detect the individual chemicals that are vaporized at very specific temperatures. A detector at the outlet measures the individual components as they pass by it, which are observed as "peaks" on a graphical plot called a chromatogram. Once all the separated components have passed through the gas chromatograph column, all the peaks are compared against pure chemical standards to confirm they are what we think they are (based on "retention time," or how long they take to pass through the column) and to understand the relationship between the amplitude of the detector signal and amount of analyte (the "response factor"). The availability of pure chemical standards that give repeatable chromatography peaks allows us to identify and calculate the concentration of each analyte peak in the original sample.

There are lots of different types of detectors, including the flame ionization detector (FID) and mass spectrometer (MS). Each has its advantages and disadvantages when detecting different types of compounds. If you are lucky enough to have a laboratory with gas chromatography capabilities, it can open up a whole new world of analysis and understanding of how terpenes behave. The sampling method and frequency will help get reproducible and actionable samples, determining everything from proper harvest timing to optimizing dosages and timing the adding of cannabis to your beer. You can even track the changes that occur throughout fermentation, packaging, and shelf life, providing endless layers of data to analyze. I would recommend utilizing resources from the likes of the Brewers Association, Master Brewers of the Americas, or American Society of Brewing Chemists (plus your local trade organizations) to learn more about aroma chemistry. Understanding aroma chemistry helps us better understand the pleasurable aromas and flavors cannabis and hops provide.

Getting reproduceable samples, let alone being able to accurately identify strains of *C. sativa*, from one farm to another can be challenging. Aside from the obvious challenges of insufficient seed banking and standards for protecting the genomic integrity of different cultivars and strains, there are also more practical challenges presented by differences in growing media and environmental and stress conditions, which all influence the terpene composition in a batch. As the cannabis industry improves its testing infrastructure and producers and consumers alike start asking for, scrutinizing, and verifying more of this type of information, the quality of information should improve dramatically. Compared to many of the infrastructure advancements made by the hops industry in the last 30 years, however, there is a considerable amount of work still to be done.

DEMYSTIFYING THE "ENTOURAGE EFFECT"

Terpenes recently became a buzz word in the marijuana industry as researchers found evidence suggesting that if humans only take individual cannabinoids, such as THC or CBD, the body has a different physiological response than if a person consumes cannabis in a more "whole plant" form. The theory behind this is that the terpenes, in conjunction with primary and secondary cannabinoids, flavonoids, and possibly many other secondary metabolites, all have a synergistic effect with one another, dubbed by the cannabis industry as the "entourage effect" (Mechoulam and Ben-Shabat 1999; Williamson 2001, 401; Russo 2011, 1353; Wagner and Ulrich-Merzenich 2009, 97).[1] A corollary of this theory is that varying concentrations of cannabinoids and other secondary metabolites can have a profound influence on the psychoactive and other pharmacological effects of cannabis. Taken a step further, companies are committing significant resources toward identifying how they can manipulate the entourage effect to make more efficacious medicines or deliver more predictable and enjoyable highs. Much of this research is cutting edge and highly proprietary; furthermore, the entourage effect has also yet to be proven true.

1 The articles cited in this section provide a high-level overview of the entourage effect theory. I recommend you read them to learn more; at this point, it's an interesting theory that shows promise, but few proven results.

It's important to understand that two different entourage effects are often described in cannabis users. The first is the "cannabinoid entourage effect," which has been demonstrated through numerous clinical studies. As we saw in chapter 5 (p. 110), *Cannabis* cannabinoids elicit their effects in humans by binding to endocannabinoid receptors in the brain. The existence of these receptors led researchers to discover that the human body produces its own "endocannabinoids" (*endo*- meaning within). It was subsequently found that certain endogenously made compounds with chemical structures related to endocannabinoids can compete with or alter the binding activity of endocannabinoids with receptors in the brain. In addition, it is possible that varying the concentration of different cannabinoids and endocannabinoids can modulate the responses of one another. Overall, this "entourage effect" was suggested as a mechanism to explain why the overall physiological response to cannabinoids ingested from "whole" preparations of plant material might differ compared to if they were ingested in a pure form (Mechoulam and Ben-Shabat 1999, 136, 139).

The second entourage effect comes from cannabis marketers, when trying to establish ways to differentiate their products, describing the "whole plant medicine/terpene entourage effect." If you've purchased legal marijuana in the last ten years, you've likely heard a budtender talk about how certain varieties of marijuana will create unique types of highs. This derives from theories in academia that expand on the original cannabinoid entourage effect to posit that plant constituents like terpenes and terpenoids can additionally manipulate our physiological response to cannabinoids (Russo 2011, 1344). The terpene entourage effect theory is informed by various studies conducted on the potential of different essential oils as pharmacologically interesting compounds. Studies show terpenes can elicit physiological responses in the body by inhaling or ingesting them; however, no study to date has shown that terpenes can modulate in a demonstrable way the effects of cannabinoids in humans (Fischedick et al. 2010, 2069; Finlay et al. 2020, 5; Santiago et al. 2019, 166).

This theory has been coopted by the recreational marijuana industry to sell and differentiate marijuana biotypes. For example, budtenders will often talk about the sedative properties of certain "Indica" strains (*C. sativa* subsp. *indica* var. *afghanica*) because of their high concentration of myrcene. Myrcene ostensibly has a sedative effect, a hypothesis based on a combination of directional studies and old wives' tales. German farmers, as the story goes, used to put hops under their pillows to promote restful sleep. Great story, but there is little clinical evidence to back it up. Many "Sativa" strains (*C. sativa* subsp. *indica* var. *indica*) also contain high concentrations of myrcene, yet they are consistently marketed as producing creative and energetic highs. I won't belabor this point because this is quickly exiting my realm of expertise, but I would encourage you to treat the "whole plant medicine/terpene entourage effect" with skepticism. It may in fact be proven correct, but at this time there is little supporting evidence for the theory.

CANNABIS AROMA COMPOUNDS: A LEXICON

The primary terpenes of *C. sativa* derive from a combination of inherited genetic traits and environmental factors. Provided growing conditions are consistent, terpene synthase enzymes will produce terpenes characteristic of a particular cannabis chemovar, and then environmental factors do the rest to produce further derivative compounds. We will focus on some of the primary terpenes created by *C. sativa* to help give better language to the distinctive, yet often hard to describe, cannabis aroma. When developing the Hemperor at New Belgium, this was a very helpful exercise to know how to target specific aromas to make sure we were getting a "true-to-type" cannabis aroma. We homed in on a key term to help us standardize "it": dank. When you ask someone to define the term *dank*, they often respond that it's something that smells like cannabis flower; in turn, cannabis is described as smelling dank. This catch-22 situation can be maddening when trying to accurately describe the differences between strains of cannabis, or even just to provide more extensive descriptive language. I will do my best to describe the primary "aroma buckets" that best make up my definition of the term dank: woody, herbal, floral, stinky, and fruity. In my opinion, every strain of cannabis has these five aroma buckets in varying amounts.

The woody aroma bucket is composed of some of the most dominant olfactory monoterpenes in cannabis: myrcene and pinene. Typically, myrcene and pinene make up the majority of essential oil from cannabis, so it makes sense most cannabis smells somewhat woody to people who have been trained in this lexicon. To me, woody aromas can range from pine sap to stripped tree bark to pencil shavings. Of course, the interplay with the herbal and floral components also generates descriptions like peppery spice, dried leaves, hay, cedar, eucalyptus, and soil.

The herbal bucket is composed of many of the dominant olfactory sesquiterpenes in cannabis, such as β-caryophyllene and humulene. This group of compounds are the second-most abundant constituents of cannabis essential oil (behind myrcene and pinene). Sesquiterpenes in the herbal bucket produce a diverse array of dried and fresh herb characteristics, which include notes of ginger, sage, basil, rosemary, oregano, hops, and fresh-cut grass. The interplay of the herbal bucket with the woody, floral, and fruity buckets helps drive the aforementioned woody aromas, plus it contributes to heavily spice-driven aromas like clove, cardamom, patchouli, and caraway seed.

The floral bucket is composed of a smattering of terpenes and terpenoids, including linalool, geraniol, and α-terpinol. While these compounds make up a small percentage of the essential oil of cannabis, they are highly aroma active and contribute significantly to the perceived aroma profile. The floral bucket contributes notes of lavender, lilac, rose, bergamot, and a general perfume-like aroma. Floral aromas interact with fruity and herbal aromas, adding additional notes of peach, apricot, vanilla, citrus fruit, and grape/wine.

There are many types of compounds that comprise the stinky bucket, but the most important are the volatile sulfur compounds. Sulfur-based volatiles run the gamut of aroma compounds (sulfides, thiols, thioesters, thioketones, and others) and are important to cannabis aroma. These compounds make up tiny fractions of a percent of the overall essential oil of the plant yet are so odor active that people can easily detect them at parts-per-trillion levels. The most common stinky odor is skunky aroma, which is mostly made up of the thiols 2-butene-1-thiol and 3-methyl-1-butanethiol. Some people describe the dank cannabis aroma almost exclusively as "skunky." Other stinky cannabis odorant descriptors include onion and garlic, cheesy (isovaleric acid), and diesel fuel or exhaust. While these aromas may sound unpleasant, in the right concentrations they accentuate floral and fruity aromas, often creating tropical fruit aromas such as passion fruit, pineapple, guava, grapefruit, and gooseberry.

Finally, the fruity bucket captures many of the oddball terpenes and terpenoids. These exist at varying concentrations in the essential oil of different *Cannabis* chemovars and are arguably what defines a particular strain or chemovar's signature scent. Compounds in the fruity bucket can include monoterpenes like limonene, which smells like orange and lime peel, and β-ocimene, which smells like underripe mango; and plant-derived aldehydes, esters, and ketones that smell like vanilla, berries, bananas, and coconuts. There are many strains that explicitly call out specific fruits in their names, such as Strawberry Diesel and Pineapple Express. The fruity bucket is typically the dominant driver of these aromas.

I posit that every *Cannabis* chemovar can produce at least one characteristic from each aroma bucket, though some tend to overexpress the characteristics of one or two. It is best to understand that it is extremely difficult to fully deconstruct the aroma of any whole plant, let alone one as complex as cannabis. Plenty try, and yet, when you smell an attempt to replicate the aroma in isolation, it is frequently missing that *je ne sais quoi*—dare I say, it's dankness? We can understand the main components that make up the cannabis aroma, but what truly defines it are the small quantities of barely measurable, or even identifiable, components. The synergistic relationship between all these components is what defines cannabis's "dank."

Table 8.1 explores a few abundant terpenes and terpenoids that are frequently encountered in cannabis. The solubility in water is important to review, which we will do in chapter 10 where we look at traditional brewing points and how to best target different cannabis aroma buckets to develop novel beers. As a quick rule of thumb, the lower the water solubility of any given terpene, the later in the brewing process you should add your cannabis material to better extract it with ethanol.

Table 8.1. Common Cannabis Terpenes and Derivatives

Name	Terpenoid Classification	Aroma Descriptors	Plants Common In	Water Solubility
β-Myrcene	Monoterpene	Woody, celery, anise, pine resin	Black pepper, mango, hops, thyme, basil	Very low
Limonene	Monoterpene	Orange, lime, pine trees	Spruce and fir trees, oranges	Low
α-Pinene, β-Pinene	Monoterpene	Woody, turpentine, pine trees, spicy	Pine, sage, eucalyptus, frankincense	Very low
Camphene	Monoterpene	Pine needles, earthy, minty, woody	Douglas fir, holy basil, nutmeg, rosemary	Very low
β-Ocimene	Monoterpene	Floral, underripe citrus, mango skin	Mint, tarragon, kumquats, mango, bergamot	Very low
Terpinene	Monoterpene	Woody, lemon, mint, medicinal	Cumin, cardamom, marjoram, cilantro	Low
Geraniol	Monoterpene alcohol	Roses, citrus, floral	Roses, coriander, grapefruit, blueberries	Medium-low
Linalool	Monoterpene alcohol	Citrus, blueberry, lavender	Lavender, rose, basil, lemon, neroli, cilantro	Medium-high
α-Terpinol	Monoterpene alcohol	Lilac, slight lemon and lime	Cardamom, lemon, grapes, dill, celery	Medium-high

Note: No attempt has been made to differentiate between enantiomers for compounds with chiral centers. Be aware that differences in perceived odors and differential effects of fragrances in terms of mood and behavior have been found to be a function of chirality.

Table 8.1. Continued

Name	Terpenoid Classification	Aroma Descriptors	Plants Common In	Water Solubility
β-Caryophyllene	Sesquiterpene	Spicy, woody, pepper, clove	Hops, caraway, oregano, basil, cinnamon	Very low
β-Farsenene	Sesquiterpene	Woody, citrus, herbaceous	Apple, orange, pummelo, noble hops	Extremely low
Humulene	Sesquiterpene	Clove, woody, peppercorn	Hops, ginseng, black pepper	Extremely low
Bisabolol	Sesquiterpene alcohol	Chamomile, citrus, generic spice	Chamomile	Low
Eucalyptol	Sesquiterpene alcohol	Eucalyptus, camphor, basil, mint	Eucalyptus, bay leaf, sage	High
β-Eudesmol	Sesquiterpene alcohol	Clove, turpentine	*Atractylodes*	Low
Guaiol	Sesquiterpene alcohol	Tea tree, rose-wood, cypress	Nutmeg, tea tree, cumin, lilac, pine	Low
Nerolidol	Sesquiterpene alcohol	Fresh tree bark, citrus, apple, rose	Ginger, jasmine, lavender	Medium-low
Caryophyllene oxide	Sesquiterpene oxide	Cedar, carrot, earthy	Hops, basil, pepper, rosemary	Low

Note: No attempt has been made to differentiate between enantiomers for compounds with chiral centers. Be aware that differences in perceived odors and differential effects of fragrances in terms of mood and behavior have been found to be a function of chirality.

9

BREWING WITH CANNABIS: USING HEMP IN THE GRAIN BILL

HEMPSEED AS A CEREAL ADJUNCT

While hemp seed has been a staple food for several millennia, it has never been documented as an ingredient in fermented alcoholic beverages. When industrial hemp laws globally began to loosen around the turn of the twenty-first century, several adventurous brewers took up hempseed as a novel ingredient to add to their beers. Admittedly, many of these early products sought to capitalize on the "edginess" of adding any cannabis ingredient, regardless of its chemistry or bioavailability. Eventually, many commercial brewers and homebrewers found hempseed to be an advantageous ingredient in beer. Hempseed is a nutritious food and useful in the brewing process for adding essential vitamins and minerals for healthy fermentations, as well as driving flavor and body in the final beer.

When talking about hempseed, there are three main products a brewer can use to create a beer: whole hempseeds; dehulled hempseeds, often referred to as hemp hearts; and pressed hempseed oil, which can come in both unrefined "extra virgin" and refined forms. Whole hempseeds are what they sound like: they are hempseeds in their natural state, which includes a thin protective shell. The shell can be eaten, but it is brittle and tends to be slightly bitter in flavor, so many food processors choose to remove it for food products. Hemp hearts are simply the hempseed with the shell removed (i.e., dehulled); this is the most common hempseed product that you will find in natural food stores and in food products. Finally, hempseed oil is the pressed oil from hempseeds that can be used for cooking and, as we will find out, novel brewing applications.

Hempseed unfortunately does not make a very useful cereal malt due to its high oil content. Some craft maltsters I've spoken to have tried to malt hempseed in the past, but none have commercialized malted hempseed products due to its poor performance in the malthouse. This makes sense given that carbohydrates make up roughly a third by weight of hempseed, far lower than the carbohydrate levels typically found in cereal malts. Most hempseed is processed to go into specialty food products, often marketed as healthy and sustainable foods for health-conscious consumers. To date, the marketing for most commercial hempseed beers tends to either focus on hempseed as a health product or as a novel product for consumers curious about cannabis.

The main forms of hempseed that are of interest to brewers are whole seeds, which can be milled, or pressed cake, which has the oil content mostly removed and can be ground into a coarse flour. Since whole hempseed contains more oil—roughly 35 percent by weight—brewers should test the upper bounds of their inclusion rates. Whole hempseed should not exceed 15 percent of the total grist, otherwise foam quality and stability of the final beer may be affected. As we will discuss later, brewing yeasts do have the ability to metabolize hempseed oil; however, adding too much whole hempseed to your grist may add too much oil for the yeast to metabolize. For those brewers looking to push more hempseed flavor into their beers, pressed hempseed cake in whole or flour form will produce a beer with superior foam quality. Pressed hempseed cake contains roughly 10 percent oil by weight and ends up having proportionally far more protein and carbohydrate by weight, 33 percent and 43 percent, respectively (Callaway 2004, 66).

While hempseed will not contribute a significant amount of fermentable extract, it does provide protein and interesting flavors that can be used advantageously by a skilled brewer. There are many common misconceptions about hempseed's flavor; it is often falsely associated with the terpenes present in cannabis inflorescences. The common refrain among brewers who use hempseed is that is produces a unique nutty flavor in beer, like raw almonds or sunflower seeds. Toasting hempseeds adds another flavor component, yielding toasted, chocolate, and coffee-like aromas as well.

Much of hempseed's flavor comes from its relatively high protein content; brewers can use this property to drive the formation of Maillard compounds in the mash tun and boil kettle, amplify esters and phenolic compounds created by many yeast strains, or both. Maillard compounds form from reactions between sugars and proteins in the presence of heat. Maillard reactions make many of the flavors we associate in cooked foods, everything from browned meats to caramels to toasted bread. Playing off of the nutty characteristics of hempseed lends well to many different styles, especially styles that focus on malt characteristics. Brown ales that lean toward nutty flavors tend to be one of the obvious styles suited to hempseed inclusion, but even beers that incorporate more delicate nut flavors, such as English mild, German Märzen, and Belgian dubbel, would pair nicely with the flavor of hempseed.

Brewers can also use hempseed additions to drive complex esters in their beer. Due to its high protein load, hempseed is a suitable adjunct for increasing the overall load of wort protein, which can be advantageous in a variety of beer styles. Initial mash temperatures lower than common saccharification temperatures allow proteinase enzymes present in the malted barley to liberate excess amino acids present in the hempseed protein. This can create "classic" esters like the isoamyl actetate common in Belgian tripels and German hefeweizens, or unconventional esters from yeasts such as kveik strains. I recommend that you pair your hempseed inclusion with another high-protein grain, such as spelt, to really drive an intense ester character.

PREPARING HEMPSEED FOR MASHING

Hempseed has many unique properties that should be considered when preparing to add it to beer. The seeds are quite small, with diameters averaging around three to four millimeters (about ⅛ inch). Hempseeds can be quite brittle and gummy, which can make their preparation challenging through traditional milling. Some brewers choose to have their hempseed pre-crushed in a dedicated mill before use. Others will simply mill the hempseed first and then use their other malts to chase and clean out any crushed hempseed that may have stuck to an auger. If you have a complicated or long-pull auger system, you may want to consider getting your hempseed pre-milled and add that grist directly to your mash tun. If by some twist of bad luck some crushed hempseed gets stuck in a dead leg of your auger system, the residual hempseed may turn rancid and affect the flavor quality of subsequent batches. If you have a short-pull auger, or a system that can be cleaned easily, these precautions are probably unnecessary.

STORING HEMPSEED

Due to hempseed's relatively high oil content, which causes oxidative instability, it is best to store the seeds cold or in airtight containers. If left in warm and wet environments for extended periods of time, such as near a brewhouse, they have the potential to go rancid and add unpleasant flavors to beer. It is best practice to purchase hempseed from a trusted vendor when you need it and to taste each lot as you receive them to screen for off-flavors. Do not store hempseed after you have toasted it, as it will stale quicker. Always toast your hempseed within a day or two of use.

How to Toast Hempseeds

Hempseed tends to have a low smoking point, so it's best to toast hempseed at a low temperature.

Oven Method: Preheat your oven to 225°F (107°C). Evenly spread your raw hempseed on a nonstick baking sheet and bake for 5–10 minutes until fragrant. If the seeds do not seem adequately toasted, shake the pan to toss the seeds and then bake another 3–5 minutes.

Skillet Method: Preheat a dry skillet over medium-low heat. Add hempseed to the skillet to make a dense but even layer. Cook for 2–3 minutes, then toss the seeds and continue toasting to cook evenly. The seeds should toast in 5–10 minutes total, depending on your desired toast level.

RECIPES USING HEMP

HEMP HARVEST GERMAN MÄRZEN

For 5 US gallons (18.9 L)

Original gravity: 1.050 (12.5°P)
Finishing gravity: 1.10 (2.5°P)
IBU: 20
Color: 8.64 SRM (17 EBC)
Brewhouse efficiency: 75%
Boil time: 75 minutes

This recipe plays off the toasty and nutty characteristics of hempseed in a clean, flavorful lager. Märzens traditionally use decoction mashing to drive the formation of melanoidins and other Maillard compounds; the extra boost of protein from the hempseed will drive these compounds even further. Do at least one decoction, but preferably two or three.

GRAIN BILL

6.17 lb. (2.8 kg) Pilsner malt
2.86 lb. (1.3 kg) Vienna malt
1.54 lb. (0.7 kg) dehulled hempseed, pressed hempseed cake, or ground hempseed flour

HOPS

0.5 oz. (14 g) German Perle (7% AA) @ 75 min.
2.8 oz. (79 g) Hallertau Mittelfrüh (3.5% AA) @ flame out – whirlpool/steep 10 min.

YEAST

White Labs WLP820 Oktoberfest/Märzen Lager Yeast

DIRECTIONS

1. Mash in with 3 gal. (11.3 L) of water at 143°F (62°C) and rest for 30 minutes.
2. Pull off 20% of wort (roughly 0.6 gal., or 2.3 L) for first decoction and boil it for 10 minutes.
3. Add decoction back to your mash and mix. This should raise the mash temperature to 156°F (69°C) depending on your mash tun's insulation. Rest for 20 minutes.
4. Pull a second decoction (20% of your volume) and boil it for 10 minutes, then add it back to the mash. (You can skip the second decoction if you wish and proceed to the sparge.)

5. Sparge with 170°F (77°C) water until you collect 6 gal. (23 L).
6. Boil for 75 minutes, adding Perle hops at the beginning of the boil.
7. Add Hallertau Mittelfrüh hops at flame out and whirlpool for 10 minutes.
8. Cool wort to 50°F (10°C), then pitch WLP820 yeast.
9. Ferment at 50°F (10°C) for 15 days, or until the gravity is stable for 7 days.
10. Cool green beer to 32°F (0°C) and rack off of the yeast.
11. Lager for 4–8 weeks before packaging and serving.

SUPER SPICE RUSTIC SAISON

For 5 US gallons (18.9 L)

Original gravity: 1.065 (16°P)
Finishing gravity: 1.006 (1.5°P)
IBU: 15
Color: 4.6 SRM (9.1 EBC)
Brewhouse efficiency: 75%
Boil time: 60 minutes

This saison recipe uses hempseed as both a flavor and a way to drive high amounts of esters and phenols. The combination of hemp and spelt should give a distinctive farmhouse quality to the beer. The combined ferulic acid and proteinase rest should allow the saison yeast to really express itself. The addition of *Brettanomyces* at packaging is optional, but it will drive further complexity in the beer as it ages.

GRAIN BILL

9.04 lb. (4.1 kg) Pilsner malt
1.1 lb. (0.5 kg) hempseed
2.42 lb. (1.1 kg) malted spelt

HOPS

0.21 oz. (6 g) German Magnum (13.5% AA) @ 60 min.
3.35 oz. (95 g) Czech Saaz (4.5% AA) @ flame out – whirlpool/steep 20 min.

YEAST

Omega OYL-500 Saisonstein®
White Labs WLP650 Brettanomyces bruxellensis
(secondary fermentation in package)

DIRECTIONS

1. Mash in at 113°F (45°C) for 15 minutes.
2. Raise the mash temperature to 149°F (65°C) for 40 minutes. Mash out at 169°F (76°C).
3. Sparge with 169°F (76°C) water until you collect 5.5 gal. (20.8 L).
4. Boil for 60 minutes, adding German Magnum hops at the beginning of boil.
5. Add Czech Saaz hops at flame out, then whirlpool for 20 minutes.
6. Cool wort to 68°F (20°C), then pitch OYL-500 yeast.
7. Ferment at 90°F (32°C) for 9 days, or until the gravity is stable for 3–5 days.

8. Cool green beer to 37°F (3°C) and age for 10 days. Rack green beer off yeast.
9. Add 100 mL of WLP650 yeast before packaging if you want to drive further earthy, funky, and spicy aromas.

BREWING NOTES

Depending on your water chemistry, you may want to lower your water pH with a little lactic acid in the mash tun and during the boil. You may also want to use a kettle fining agent to help precipitate out protein prior to cooling your wort.

FORTIFICATION: HEMPSEED ABBEY ALE

For 5 US gallons (18.9 L)

Original gravity: 1.070 (17°P)
Finishing gravity: 1.014 (3.5°P)
IBU: 30
Color: 25 SRM (49.3 EBC)
Brewhouse efficiency: 72%
Boil time: 180 minutes (3 hours)

Belgian monks make Trappist ales to fund the mission of their monasteries and to make a hearty, nutritious beer to fortify themselves during their Lenten fasts. Considering how nutritious hempseed is, perhaps our monastic brothers would consider adding it to their next beer! For this Belgian-style dubbel, the recipe uses both the nuttiness from toasted hempseed and selective mashing techniques to drive esters and phenols. If you are toasting the hempseed yourself, you may want to try toasting it a touch longer than you normally would. The extra toast can help drive some chocolate and coffee aromas in the hempseed, adding to the depth and complexity of this beer.

GRAIN BILL

8.37 lb. (3.8 kg) pale malt
2.64 lb. (1.2 kg) melanoidin malt
0.66 lb. (0.3 kg) heavy toasted hemp seeds
1.32 lb. (0.6 kg) Munich malt
0.22 lb. (0.1 kg) chocolate malt

HOPS

0.53 oz. (15 g) German Magnum (13.5% AA) @ 60 min.
2.93 oz. (83 g) Belgian Strisselspalt (2.1% AA) @ flame out
 – whirlpool/steep 20 min.

YEAST

White Labs WLP500 Monastery Ale

DIRECTIONS

1. Mash in at 122°F (50°C) for 10 minutes.
2. Raise mash temperature to 153°F (67°C) and rest for 30 minutes.
3. For extra credit, you can do a double decoction mash to raise from the strike temperature to the mash rest temperature and raise it again to the mash out temperature. This will help drive the formation of more Maillard compounds.

4. Sparge with 169°F (76°C) water until you collect 6.6 gal. (25 L) of wort.
5. Boil for three hours. The long boil is to drive further Maillard reactions.
6. Add the German Magnum hops with one hour left in the boil.
7. Add the Belgian (or French) Strisselspalt hops at flame out and whirlpool for 20 minutes
8. Cool wort to 59°F (15°C), then pitch WLP500 yeast.
9. Ferment at 65°F (18°C) for a maltier beer, or at 75°F (24°C) for a more estery beer.
10. Ferment for 7 days or until the gravity is stable for 3–5 days.
11. Cool the green beer to 32°F (0°C), hold for 3 days, then rack beer off the yeast into a secondary fermentor. Lager in secondary for at least 2 weeks.
12. Package the beer with high carbonation, ideally 3.5 volumes of carbon dioxide. Extra credit for bottle conditioning.

HEMPSEED OIL AND YEAST HEALTH AND VIABILITY

Pressed hempseed oil also presents a potentially novel ingredient for brewers as a yeast processing aid. Normally, brewers must aerate their wort with compressed air or pure oxygen to trigger yeast growth in their fermentation. Yeast cells need to produce unsaturated fatty acids to create new cells and take up oxygen to create these fatty acids. During the initial lag phase of fermentation, yeast cells come out of dormancy and will take up dissolved oxygen in the wort to begin reproducing. During this time, the cells must rely on their internal energy reserves, glycogen and trehalose, to begin their metabolic activity. Over the course of a normal fermentation, yeast cells will deplete these energy reserves first and then turn their attention to wort sugars to continue fermentation. As fermentation winds down, the cells will replenish their glycogen and trehalose reserves with some of the remaining sugar left in the now green beer, but they cannot replenish their reserves of unsaturated fatty acids because the beer is devoid of these. Thus, if brewers wish to repitch their yeast into a subsequent fermentation, they must store their yeast cold to make sure the yeast goes dormant for a short period of time. When they repitch the stored yeast, the brewer must add supplemental oxygen to the new batch of wort to start yeast metabolism again.

Dissolved oxygen added at any point in the beer-making process has the potential to create compounds that will lead to oxidation and premature staling of the beer. Commercial brewers spend inordinate amounts of time devising ways to minimize dissolved oxygen in beer and significant research and engineering has gone into eliminating as much oxygen ingress from the beer-making process as possible. Adding oxygen to wort is, at best, considered a "grand bargain" between creating healthy fermentation conditions initially while sacrificing some amount of the product's shelf life. Staling and oxidation is a tricky subject: eliminating oxygen ingress at one point of the process may not result in the elimination of staling compounds in your finished product if you have ingress elsewhere. Studying the minimization of oxidation requires a holistic perspective of every step of the brewing process to see definitive results.

It was through this holistic quest to minimize dissolved oxygen in the brewing process that researchers in the early 2000s came up with the idea that, instead of adding supplemental oxygen to wort, they could directly add unsaturated fatty acids to brewer's yeast to see if that resulted in successful fermentation (Moonjai et al. 2002, 227). The experiment demonstrated that adding unsaturated fatty acids—in the 2002 study the researchers used linoleic acid—to yeast in a strictly anaerobic environment allowed the yeast to produce an acceptable beer with similar characteristics to a traditionally aerated beer. At New Belgium, Grady Hull (my old boss and mentor) conducted his master's thesis by adding olive oil to the brewery's stored yeast and conducted successful fermentations of Fat Tire. In his study, Grady also concluded, using New Belgium's trained sensory panel, that the Fat Tire batches supplemented with olive oil showed fewer markers for oxidation throughout their shelf life when compared to control batches of Fat Tire (Hull 2008, 22).

Grady chose olive oil for his study because olive oil contains high levels of linoleic acid and is readily available and relatively cheap compared to the synthesized linoleic acid used in the Moonjai et al. study. Hempseed oil contains 93 percent unsaturated fatty acids by weight, whereas olive oil contains only 84 percent by weight (Callaway 2004, 66). Theoretically, this makes hempseed oil a more effective raw material by weight for this purpose.

You may be wondering to yourself, "Couldn't I just mash in with sufficient amounts of hempseed and extract both the flavor and the unsaturated fatty acids?" While it is tempting to say yes, the more responsible answer is no. While you certainly will carry some quantity of essential fatty acids into your finished wort, there is too much variability in the brewing process from batch to batch to make a high-quality calculation of total unsaturated fatty acids in solution. The precise control of unsaturated fatty acid supplementation to pitched yeast is crucial for a successful fermentation, just as it is with controlling pitch rate, temperature, and nutrition. Too much variability in the process will result in unacceptable amounts of wasted beer and hemp.

I was hoping to have completed an in-depth trial to study the efficacy of hempseed oil as an unsaturated fatty acid supplement for yeast but, due to the

COVID-19 pandemic, I have not been able to conduct it as of writing this. I do hope that hempseed oil as a yeast supplement will be considered for future research as it could be an interesting value-added ingredient for improving yeast health and beer quality. The idea is theoretical at this point; however, there are signs it could be a fruitful area of study. The 2002 study by Moonjai et al., as well as Grady Hull's research at New Belgium, suggest that yeast's ability to utilize a broad spectrum of unsaturated fatty acids not only yields a successful fermentation, but also raises the possibility of greater shelf stability. The demand for high-quality hempseed oil is low at present, so it will not be a cost-effective strategy to implement on a large scale anytime soon, but as the industrial hemp market grows there is potential for costs to fall. In the meantime, it is a fun thought experiment and, I hope, a topic that researchers will explore in the near future.

Dosing Hempseed Oil to Replace Aeration

I recommend trying a variety of dosage rates of hempseed oil to dial in your fermentation performance. Higher dosages may be required for high-gravity beers; likewise, lower dosages will successfully ferment low-gravity beers. There are recommended levels to start with in the steps below. I recommend doing a small-batch fermentation as a test prior to scaling up this method—there will likely be many unique considerations for your brewery that will not be captured in this method.

Instructions:

1. Harvest your yeast as you would normally.
2. Measure the volume of your harvested yeast slurry and stir the batch to homogenize your cells.
3. Count your total yeast in suspension via hemocytometer, or other appropriate instrumentation. Multiply your cell count of total cells per milliliter by the volume harvested (in milliliters) for your total yeast cell count.
4. Calculate your pitch rate for the total cells needed.
5. Per Hull (2008) and Moonjai et al. (2002), add 1 mg hempseed oil per 25–75 billion cells pitched. For higher gravities, test closer to 1 mg per 25 billion cells. For lower gravities, test 1 mg per 50–75 billion cells. Calculate your total hempseed oil by the following formula:

$$\text{total hempseed oil required (in mg)} = \frac{\text{total yeast cell count}}{\text{target yeast concentration per mg of hempseed oil dosed (in billions cells)}}$$

6. Add your calculated total hempseed oil to your yeast brink and mix thoroughly. Allow to sit for 24 hours for the yeast to fully absorb.
7. Pitch yeast as normal into a new fermentation.

10

BREWING WITH CANNABIS: HEMP AS A FLAVOR

We discussed hemp's utility as a cereal adjunct in chapter 9. Now we will focus on hemp's potential as a flavor and aroma contributor for beer. The focus of this review will be on hemp as a flavoring for commercial beer; however, this knowledge is also relevant for homebrewers looking to flavor their beer with either hemp or marijuana, where they are legally able to do so. In theory, any part of the hemp plant that contains glandular trichomes has the potential to be a flavoring ingredient, including the stems and leaves, but the trimmed flowers will provide the best flavor quality.

There are myriad ways to infuse aromatic compounds into beer. We must look at the hemp plant holistically to identify the components and brewing processes that yield optimal results for flavor. Due to the similarities of *Cannabis* with *Humulus*, the logical starting point is to evaluate hemp's performance at traditional hopping points: hot-side additions, focusing on the kettle, whirlpool, and hop back; cold-side additions that employ dry hopping both during and after fermentation; and as a finishing flavoring addition to cask-conditioned ales or as a liquid concentrate added at filtration. When navigating this chapter, know that we will review the traditional hops addition points and create a quick rubric for how a brewer might exploit the aromatic "goods" of the cannabis plant.

Hemp inflorescences (hemp "flower") can take many suitable forms for brewing and are similar in performance to comparable hop products: whole flower, pelletized flower, extracted concentrate, and distilled essential oil. All these hemp products have applications in the brewing process, but all have crucial drawbacks in certain cases. Chapters 6 and 7 looked at strategies to solubilize cannabinoids and dose such products into beer, and which addition point is most advantageous for the brewer. In this chapter we will look at getting the odor- and flavor-active compounds of cannabis essential oil, previously discussed in chapter 8, into your beer. It may help to refer to that chapter again, including table 8.1 (pp. 173–4), as you read this present discussion. The bewilderingly diverse components that comprise cannabis essential oil all behave differently when added at various points of the brewing process. We can helpfully break up the mechanisms of aroma/flavor extraction during the brewing process into several areas: hot versus cold extraction, water-based versus alcoholic extraction, and the biological action of yeast during fermentation affecting aroma extraction and expression. Keep in mind, this review is based on our understanding of hops aroma chemistry and, thankfully, this knowledge can be directly translated into using cannabis for the same purpose in brewing.

We will explore four main classes of aroma compounds that are frequently cited in hop research and are critical for extracting aromas from cannabis: terpenes, terpenoids (oxygenated terpene derivatives), sulfur-containing compounds, and plant-derived esters. Each of these have certain chemical properties that will determine the optimal dosing point for highlighting their aroma characteristics; however, in general, the beer making process will extract all of them to varying degrees, although, depending on the process, not much of the desired aroma compounds may survive into the final beer. It is easy to get lost in the weeds when trying to prioritize one compound or a set of compounds over the other when thinking about developing these flavors. For example, brewers often talk about whirlpool hopping or mid-fermentation dry hopping when creating juicy New England-style IPAs, but they don't think about how late-fermentation dry hopping may influence the perception of "juicy" aromas: prioritizing only mid-fermentation dry hopping may coax out more compounds that are beneficial to the desired juicy aromas but it also suppresses other aromatic compounds that help amplify them. Ultimately, by dialing in just one type of addition to maximize extraction at that stage, the overall aroma profile may fall short and the brewer will be disappointed in the result. Similarly, we must think about which attributes of cannabis are pleasing to us and devise strategies to make sure we are maximizing these aromas. At New Belgium, when we were developing the Hemperor, we made a conscious decision to drive aromas that would be recognizable as a bag of cannabis flower. We defined those aromas as a mix of herbal, woody, floral, diesel, and skunky, which we all agreed constituted the essential traits of certain *Cannabis* varieties. We then made decisions about how to brew the Hemperor in such a way that would maximize the overall perception of the "true-to-type" cannabis aroma. During our development, we explored multiple addition points to understand how the aromas could be manipulated to this end—in many cases, we found that the timing of our addition was critical.

Unfortunately, the cannabis industry has yet to catch up to the hop industry in many respects. First, there is no standardized naming convention for individual *Cannabis* varieties and cultivars for both the hemp and marijuana industries. Naming of hemp cultivars follows some level of standardization, but it is often up to the grower or even retailer to name them. The situation with marijuana strains is far more nebulous; indeed, the international cultivated plant code does not recognize any varietal marijuana crop as a cultivar, which is why they are referred to as "strains" by most botanists (Small 2015, 311). The situation is improving with every year; however, the industry still has some old habits held over from its clandestine past, where growers and breeders kept little documentation, cross-bred strains frequently, and shared information via word of mouth. This makes it difficult for us as brewers and consumers to get standardized, repeatable data that we can share. For now, I would encourage you to view each of these cultivars and strains in a microcosm: if you like the aroma qualities of a certain varietal crop from a certain grower, build upon that relationship. Use a combination of sensory evaluation along with laboratory analysis to help you develop your own aroma attribute lexicon that you can use in developing your beers. Another grower may claim to have the same biotype or chemovar for a cheaper price, but that does not necessarily mean that it will turn out the same in your beer. I hope that, as time moves on, genetics and growing practices can be verified and standardized, but the market currently is hit and miss. We will discuss more quality assurance strategies in chapter 11.

This also speaks nothing of terroir. As we know from chapter 8, growing conditions, climate, and soil composition are critical factors affecting the expression of aroma compounds in cannabis. I view this as a net benefit, as individual growing regions may be able to differentiate themselves by creating unique properties only found within their microclimate. This could in turn create cannabis beer and beverages more akin to wine varietals, where regional expressions are highly prized and sought after. We are a long way off from that, but I think it is likely to be a driver of growth and innovation in the future.

HEMP AROMA PRODUCTS AVAILABLE

Whole Inflorescence ("Flower")

Whole *Cannabis* inflorescence ("flower") is the least processed form of the cannabis plant that is of interest to brewers for the purposes of aroma. Whole flowers will often be trimmed of excess fan and sugar leaves, although some growers will simply dry the flowers and sell them unprocessed. Too many sugar and fan leaves on the flower will lead to excess grassy, hay, and vegetal aromas and will dilute the concentration of glandular trichomes in the sample. It is best to specify some level of trimming with your vendor to make sure you have the right ratio of vegetation to trichomes. When brewing with whole flower, it tends to extract and behave similarly to whole-cone hops. Compared to pellets, whole flower has less surface area, so the extraction efficiency is reduced as compared to pellets. For smaller operations, it may be advantageous to grind or powderize the whole flower prior to use to increase the extraction efficiency. It is best to store whole flowers in their natural state, ideally away from excess heat, light, or oxygen.

Pelletized Flower

Pelletized cannabis flower will more than likely come from a vendor that processes both hops and cannabis; at this moment in time, cannabis in this context means industrial hemp only. After the hemp flowers are dried and trimmed, they are fed into a hammer mill and ground to a uniform consistency prior to being put through a pellet mill, or press. While pellets are now the standard form for hops to come in, they are still rare in the cannabis industry. As cannabis becomes more mainstream in brewing, I imagine pellets will be a more common product.

Steam-Distilled Essential Oil

Steam-distilled essential oil is most often sold by industrial hemp operations that are trying to maximize their yields of both essential oil and cannabinoids for specialized vaporizer products and CBD concentrates. The farm will cut down the cannabis plants and immediately add the biomass into a steam distillation apparatus, where steam will volatilize the aromatics, separating them from the biomass. That aroma-rich steam is then condensed and de-watered to make a highly concentrated, true-to-type aroma product. Typically, the farm will then take the residual biomass, dry it, and use solvent extraction to isolate the cannabinoids. At the end of the cannabinoid extraction and concentration process, the steam-distilled essential oil is blended back into the cannabinoid concentrate and sold as vaporizer cartridges, dabs, or other smokable products. Steam-distilled cannabis essential oils present the same opportunities to brewers as steam-distilled hop essential oils: they are highly concentrated, easy to handle, and minimize process losses in the brewery. Because steam-distilled oils are highly processed and concentrated, there are limitations to using them for creating true-to-type aroma profiles, but they can be useful for boosting certain aromas.

Extracts

Cannabis extracts come from the cannabinoid concentration process outlined in chapter 6; however, in the extract production process, rather than being further processed into a water-soluble format after solvent extraction, the extract is provided in its current “crude oil” form or in the slightly more refined “winterized” form (see p. 130).

The extraction process concentrates both cannabinoids and terpenes, but, due to the high volatility of monoterpenes and some sesquiterpenes and diterpenes, the extraction process often volatilizes many terpenes from the extract crucial to aroma. Cannabis extracts are similar to the hop extracts made from liquid carbon dioxide extraction used in brewing for bittering and sometimes aroma in the brewhouse. Cannabis extracts may be appropriate if you are targeting specific aroma compounds that are stable enough to survive and be concentrated in the extraction process.

Constructed Flavors

Constructed cannabis flavors come from the flavor and fragrance industry and are also referred to as “natural flavors.” Flavor companies will analyze the individual chemical components of cannabis aroma and will build a cannabis-like aroma or flavor product using components from other ingredients. For example, in a cannabis-type flavoring, the linalool fraction of the flavoring would not be derived from the cannabis plant, but rather coriander or oranges. Constructed flavors can be useful when formulating cannabis beers when regulatory restrictions do not allow you to use actual cannabis material. These flavorings have varying degrees of fidelity to the true-to-type aroma of cannabis depending on the formulation. Constructed flavors are often customizable and can be tuned to the properties that you are looking for.

ADDING HEMP AROMA PRODUCTS IN THE BREWING PROCESS

Wort Boil

The kettle is the first point of addition for most aroma products not part of the grist. Hops and spices are, conventionally, the aroma products that brewers add to the kettle. Based on our knowledge of how these ingredients behave in wort, we can evaluate how cannabis fits into this process. Thinking back to our mechanisms of aroma/flavor extraction listed earlier, the kettle is a hot, water-based extraction method.

What we put into the kettle will impact downstream brewhouse and cellar processes. Aside from the extraction of beneficial and desirable aroma compounds from exogenous ingredients added to the wort, boiling wort accomplishes eight additional goals:

- Volatilization of unwanted malt- and hop-derived aromas such as DMS from malt and excess grassy and vegetal aromas from hops
- Isomerization of hop alpha acids to drive bitterness
- Precipitation of protein and polyphenol complexes that can produce undesirable haze in the final beer and cause fermentation stress
- Concentration of wort sugars to hit target original gravities
- Chelation of metals and minerals such as calcium oxalate to prevent downstream beer stone formation
- Sterilization of wort to prepare it for fermentation
- Formation of color and flavor components through the Maillard reaction
- Reduction of pH to set optimal conditions for fermentation

Boiling wort with no hops will still accomplish several of these goals; however, hops are essential for alpha acid isomerization, and play a critical role in precipitation, chelation, and pH reduction. Without adding some amount of hops at the beginning of the boil, or other processing aids to accomplish the same goals, most brewers would produce beer of dramatically lower quality. Through these critical roles, we can evaluate hemp's potential to supplant hops in wort boiling operations.

To cut a long story short, there's a reason why hops are used at the beginning of the boil. Since *Cannabis* lacks the alpha acids responsible for providing modern beer's characteristic bitterness, there is little reason to overly analyze the other potential contributions from adding cannabis at the beginning of the boil. We add hops for the bitterness, the other benefits just come as a bonus. If we did not want bitterness in our beer, there are plenty of other additives that would aid in precipitation, chelation, and pH reduction far better than an expensive cannabis product.

While some components of cannabis glandular trichomes do contribute to perceived bitterness, there is no component so identified that can provide a suitable substitute to isomerized alpha acids from hops. I have heard many anecdotal accounts that cannabinoids contain some degree of bitterness; however, adding cannabinoids at the beginning of the boil for bitterness is costly and inefficient. Best stick to hops.

Cannabinoids can undergo decarboxylation when heated (see chap. 5). This does not, however, solve the solubility issue with cannabinoids: they are hydrophobic so they will not readily dissolve in a water-based medium such as wort. Unlike hops, where the isomerization of alpha acids makes them more water soluble, there is no such advantageous change with cannabinoids. Although the boil kettle does provide enough heat to decarboxylate cannabinoids—especially when boiled for long periods of time—it produces variable results in the beer due to this solubility problem.

Whirlpool: Aroma Addition

Cannabis contributes far more significantly to overall flavor and beer quality if added at the end of boil, either in the kettle, whirlpool, or hop back. As mentioned at the start of this chapter, we will focus on hemp, but know that the same principles discussed here apply to marijuana and can be investigated at home in jurisdictions where marijuana is legal.

Hemp contains similar concentrations of essential oils to hops and, in certain cases, even higher concentrations. Hemp behaves similarly to hops in the whirlpool and, as you might expect, the form it comes in makes a difference. Whole flower contributes different qualities to beer than pelletized hemp, which is mostly a function of surface area. When adding hemp in the whirlpool, targeting various types of aroma compounds allows you to identify which processes will favor those compounds that are present in the essential oils.

Whirlpooling is still a hot, water-based extraction method, though it strips fewer volatile compounds than active boiling. The hydrophobicity and volatility of terpenes in general is still a major consideration in the whirlpool. Despite the relatively high concentration of monoterpenes (e.g., myrcene or α-pinene) in the essential oil, these show low solubility in the wort during whirlpool. Monoterpene aromas from exclusively whirlpooled hemp additions will be comparatively less intense than in beers that have been dry hemped, not only due to the heat that flashes off these volatiles in the form of pleasant-smelling steam but also because there is no ethanol present to help solubilize the terpenes. At New Belgium, we frequently test the terpene content of beers produced with only whirlpool hopping versus beers that were exclusively dry hopped and it is not uncommon to see mono- and sesquiterpenes in the whirlpooled beers at 50% of the concentration found in the dry-hopped versions. The same principle applies in the case of hemp terpenes.

Whirlpooling does have the ability to catalyze oxidative reactions and transform terpenes into their oxidized derivatives. For example, the sesquiterpene β-caryophyllene, which is produced in abundance in many hemp cultivars, can be oxidized in the whirlpool into the sesquiterpenoid caryophyllene oxide, a terpenoid frequently studied for its role in creating the classic peppery spice character traditionally derived from European noble hops.

In fact, the whirlpool is arguably the optimal process for the extraction of terpenoids. Compared to terpenes, the oxygenated functional groups in many terpenoids confer a relatively high solubility in water-based solutions and terpenoids generally have higher boiling points, allowing them to be easily extracted and stay in solution in the hot, water-based whirlpool. Some terpenoids can exist in a free form, or bound to a sugar molecule as a terpenoid glycoside. By binding various classes of compounds to sugars, living things use the resulting glycosides to store or transport useful compounds, enzymatically hydrolyzing the sugar to liberate the stored compound when needed. The heat from the whirlpool can partially liberate terpenoids from terpenoid glycosides, amplifying the aroma intensity of the wort. What terpenoid glycosides survive the boil and whirlpool can be subsequently hydrolyzed by yeast, imparting additional aroma intensity to the actively fermenting beer. Yeast hydrolysis of bound glycosides is one of the characteristics of yeast biotransformation, a subject we will get to shortly.

Terpenes and terpenoids are not the only aroma constituents of hemp and hop essential oils; both plant-derived oils also contain several other aroma active constituents that are of interest to brewers. While they typically exist at low concentrations, many of these constituents are highly aroma active, perceptible in the parts per trillion range. They can be challenging to study in the context of brewing because many of these compounds are volatile, present in very low concentrations, and difficult to measure. Arguably, these compounds differentiate cannabis cultivars and strains from one another and provide their signature aromas. While we may not fully understand how they behave in the brewing process yet, we do know they play an essential role in aroma expression. The two subcategories of this group worth focusing on are sulfur-based aroma compounds (e.g., sulfides, thiols, thioesters, and thioketones) and esters. These two categories often provide many of the complex fruit aromas key to hemp and hops and are generally regarded as drivers of appealing aromas in beer.

The whirlpool plays a complex role in the expression of sulfurs and esters, both for its role in solubilizing them and for selectively volatizing aroma compounds like monoterpenes that might otherwise mask our perception of sulfur and ester-based aromas. The heat of the whirlpool is a net detractor for many volatile sulfur compounds and esters if they exist in a free form. Sulfurs such as polyfunctional thiols (a thiol that also contains another functional group) also tend to oxidize quickly, so the splashing of kettle operations combined with wort aeration more than likely reduces their aromatic intensity. Thiols can exist as glycosides that can be hydrolyzed downstream during fermentation, so the whirlpool could be an appropriate place to extract these glycosidic compounds for later release in beer (Roland, Delech, and Dagan 2017, 171). Additionally, many plant-derived esters can also be extracted in the whirlpool that can in turn be further transesterified by yeast into other novel esters, or converted into alcohols, ketones, and other aroma compounds (Steyer et al. 2017, 137).[1]

Brewing engineers designed whirlpools to handle solid, vegetative material so I would recommend that you focus your whirlpooling efforts on the hemp products that contain vegetative material: mostly whole hemp flower and hemp pellets. Products like distilled oil provide greater value in cold applications and tend to be highly concentrated; the heat from the whirlpool will volatilize the aroma compounds from these highly concentrated products, making them less economic at this addition point. Whole hemp flower is a great material to consider, do make sure to grind down the flower into at least a coarse powder to increase the surface area for more efficient extraction. Pellets are useful since the pelletizer has already done the work for you, making uniform pellets of small particles that will readily disperse under hot conditions.

---

[1] I would suggest reading some of the cited articles, they highlight very interesting developments in our knowledge of beer aroma. I have also found "The contribution of geraniol metabolism to the citrus flavor of beer" (Takoi et al. 2010) and "The freshening power of Centennial hops" (Kirkendall, Mitchell, and Chadwick 2018) particularly useful in my studies of hop aroma. These are included in the bibliography.

Mid-fermentation Dry Hemp

The process of adding hemp during an active fermentation is informed by new styles of IPA that use hop additions during active fermentations, such as the New England IPA style. Brewers began using mid-fermentation dry hopping techniques based on word-of-mouth recommendations from their fellow brewers and experimented with new (at the time) tropical and fruity hop varieties such as Citra, Mosaic, and Galaxy. It is no coincidence these hops are not only rich in many common terpenes, but they also have above-average concentrations of sulfur-based aroma compounds such as thiols and thioesters. Brewers reported anecdotally that beers produced in this way tended to have appealing, fruit-like characteristics that many simply described as "juicy." The Brewer's Association now even has a whole stylistic category in its style guidelines, which is a testament to how popular this processing technique is with brewers. Fortunately, the academic community has also begun studying these processing techniques and has identified several biochemical processes that underpin the anecdotal reports by brewers.

While most of the popularity of mid-fermentation dry hopping is, of course, rooted in the tradition of making hoppy beers, I argue it is equally as important for hemp and marijuana when formulating beers. Many cannabis cultivars and strains share aroma and name similarities to fruits and, arguably, these characteristics are a significant driver of purchase intent among consumers. Strains such as "Pineapple Express," "Forbidden Fruit," and "Super Lemon Haze" all capture the attention of cannabis consumers because of their pungent and unique fruity characteristics; it is logical that any crossover product in beer should seek to amplify these types of aromas.

Thus, mid-fermentation dry hemping is the optimal process for developing modern, fruity cannabis aromas in beer. The process itself is a (relatively) cold extraction method that features changing concentrations of ethanol (based on timing) in the bulk solvent and utilizes active yeast fermentation to create many characteristic aromas. Of all four processes outlined in this chapter, it has the highest number of variables to consider. You may think that, as a "warm" cellaring operation, mid-fermentation dry hemping will produce many of the same characteristics as whirlpool hemping, but you must also consider the longer extraction time at these cooler temperatures. Adding hemp during fermentation allows for increased contact time, increasing the extraction of hemp aroma compounds. Brewers must also weigh several other variables:

- Stripping of volatile terpenes through carbon dioxide formed during fermentation and vented from the tank
- Binding, absorption, or partitioning of volatile compounds to foam and yeast cell walls
- Changes in solubility of certain compounds depending on ethanol concentration
- Overall solubility of hydrophobic aroma compounds like monoterpenes
- Biotransformation of hemp aroma compounds by yeast
- Liberation of glycosidically bound aroma compounds
- Scavenging of oxygen by yeast
- Hemp creep and excess ethanol formation, similar to hop creep

These variables do not exist in isolation; they all occur simultaneously, so there's a bit of wiggle room and still a lot of uncertainty as to how the variables interact and influence the final beer's aroma. It is advisable not to add any hemp, or hops for that matter, until your pitched, oxygenated yeast undergoes its logarithmic growth phase and consumes all of the dissolved oxygen. This helps get the yeast healthy and happy, which is a crucial component for successful biotransformation later in the process, not to mention critical for a successful fermentation. Adding resins and essential oils is inherently stressful on yeast, so give the little buggers a break while they are getting started. Furthermore, introducing hemp in the presence of oxygen increases the potential for free thiols and other aroma compounds to be oxidized and their aromas will not carry over into your beer.

The first potential mid-fermentation addition point will occur when the yeast is vigorously fermenting after the beer is roughly 30–50 percent attenuated. At this point of fermentation, there is a low concentration of ethanol and the beer can, for all intents and purposes, be considered a water-based extraction medium. Bear in mind that yeast produces high volumes of carbon dioxide that can strip many of the volatile components (terpenes and non-glycosidic sulfurs or thiols), so this addition point is best for selective extraction of stable compounds like terpenoids. It generally provides less overall hemp intensity relative to later dry-hemp addition points. Since the yeast is very metabolically active, you will benefit both from the biotransformation of hemp aroma compounds and the liberation of bound glycosides present in hemp.

It's useful to define biotransformation, as it is currently one of the most buzzworthy terms in brewing, yet most definitions of the term tend to be a catch-all bucket for any manipulation of hops aroma compounds during fermentation. I've deliberately left out stripping, binding, and liberating glycosidically bound compounds from my definition, but many others consider these elements part of biotransformation. In my definition, **biotransformation** is the process where yeast metabolizes exogenously added compounds, transforming them into different compounds.

Our understanding of biotransformation mostly comes from studying how hops compounds are affected; however, compounds in spices, fruits, and, of course, cannabis are all subject to this biotransformation as well. Yeast genetics also play a crucial role, where certain yeast strains will readily carry out biotransformation processes of some kind that others will not. Most of the existing research suggests two key groups of aroma molecules that undergo biotransformation: terpenoids and plant-derived esters. This makes sense considering the two main groups of aroma compounds produced by yeast are higher alcohols and esters; naturally they would also take up exogenous sources of similar compounds and further manipulate them during yeast metabolism.

Arguably the best studied biotransformation pathway is the manipulation of the monoterpene alcohols geraniol and linalool. (Remember that monoterpene alcohols, being oxygenated derivatives of terpenes, are terpenoids.) First studied in beer by King and Dickinson (2003) and later expounded

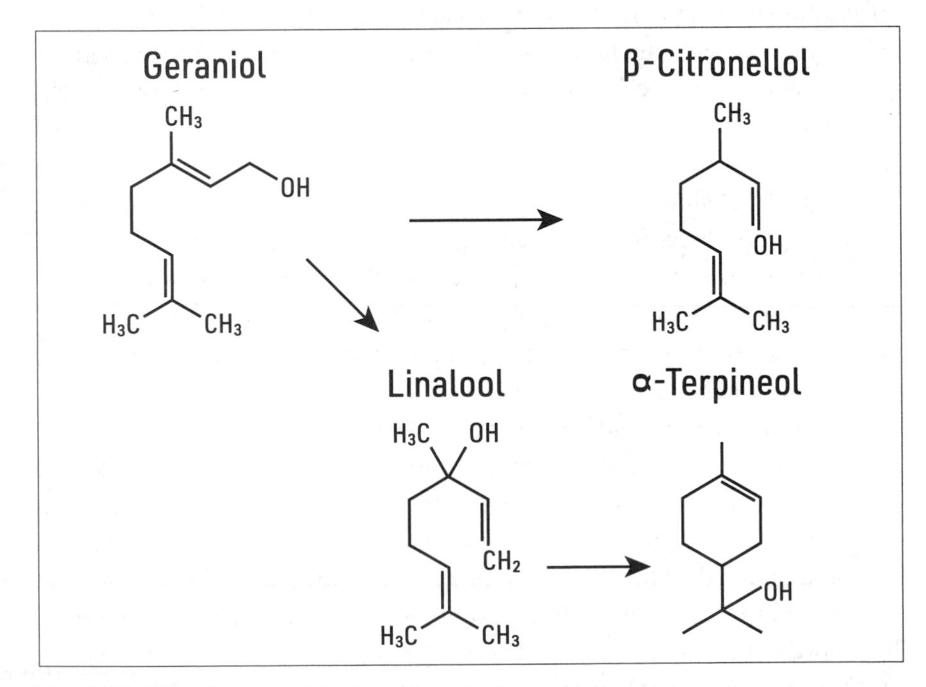

Figure 10.1. Metabolism of geraniol by brewer's yeast.

by Takoi et al. (2010), it has been shown that multiple yeast species can use this pathway to metabolize geraniol (rose aroma) and transform it into both β-citronellol (lemon-lime aroma) and linalool (Froot Loops® or lavender aroma); linalool can further be transformed into α-terpineol, which has a lilac aroma (Takoi et al. 2010, 252; King and Dickinson 2003, 56).

It was additionally indicated that yeast may be able to esterify these terpenoid alcohols into their corresponding acetate esters (King and Dickinson 2003, 59). Further research by Dr. Takoi's group at Sapporo confirmed that the progress of these reactions throughout fermentation can be observed, and illustrated how they can be subtly manipulated by brewers (Takoi et al. 2016, 91). Understanding the relative abundances of these terpenoids in materials like hemp and hops, we can predict the way yeast will biotransform these compounds into their successive forms during fermentation. For example, we should see concentrations of geraniol fall as corresponding concentrations of β-citronellol rise (Takoi et al. 2016, 88). Depending on the reaction time, we should additionally see geraniol levels be replenished due to the yeast liberating glycosidically bound geraniol in solution (Takoi et al. 2016, 86). Brewers can somewhat control the rates of these reactions based off the timing of their addition: for example, adding hemp material rich in geraniol earlier in fermentation will result in higher concentrations of β-citronellol than if the hemp were added later in fermentation.

As you dry hemp later in the fermentation process (but before fermentation is complete), the results begin to taste like beers that are traditionally dry hopped. At this stage, there is far less metabolic activity from yeast, limiting the amount of stripping of aroma-active volatiles by carbon dioxide. Additionally, due to the relatively higher concentration of ethanol, the overall solubility of hydrophobic components like terpenes increases significantly. In fact, the increase in ethanol will aid the solubility of most hemp aroma constituents.

Later in fermentation as the yeast starts entering its dormancy phase due to lack of food, its ability to biotransform hemp compounds will be greatly diminished, as well as its ability to liberate glycosidically bound terpenoids. Additionally, you should consider the stress on the yeast by adding hemp (or hops) that late in the process. If you have a sluggish fermentation with poor yeast health to begin with, such a shock may prevent you from successfully reducing diacetyl, or you may produce other yeast stress compounds such as acetaldehyde. If you have a healthy fermentation with plenty of viable yeast this will likely not be an issue. As yeast enters dormancy, it does have one last trick up its sleeve: yeast can adsorb certain monoterpenes such as β-myrcene or β-pinene as it flocculates out of beer. This technique, which brewers can also take advantage of in post-fermentation dry hemping, can help reduce the aroma intensity of these abundant monoterpenes, allowing for lower-aroma threshold odorants such as sulfurs and esters to become more perceptible in your finished beer.

One additional variable of note, although not directly related to aroma, is that hemp appears to have similar levels of amyloglucosidase activity to hops when added during an active fermentation. Amyloglucosidase is one of the enzymes responsible for releasing the fermentable sugar glucose from high-molecular-weight starches and, when active in a fermentation, will lead to overattenuation of your beer and increased ethanol levels (Kirkpatrick and Shellhammer 2018, 9121). Through multiple trials conducted at New Belgium, we found strong evidence that hemp flower added during an active fermentation will lead to overattenuation and increased ethanol production. This evidence has not been conclusively determined in an academic study; however, the results observed at New Belgium indicate that hemp shares similar amyloglucosidase activity to hops.

The mid-fermentation hemp addition point is suited to a few cannabis products, especially those that contain a diversity of aroma compounds and their glycosidically bound precursors. Whole flower hemp is always the obvious choice; however, it should be ground to a coarse powder to increase extraction efficiency. Pelletized hemp flower accomplishes the same goals as whole hemp flower but adds greater surface area to increase extraction efficiency without the brewer having to do it. Distilled essential oil products can be used in a mid-fermentation addition, but you should analyze what exactly is in your essential oil before making that determination. If the essential oil is rich in terpenoids and you are looking to highlight those characteristics, it will be a welcome addition for a mid-fermentation application. If the essential oil content is more heavily weighted toward mono- and sesquiterpenes, it may not be the best addition point due to their relatively low solubility and high volatility. Essential oils tend to be very expensive; it would be a waste of money to volatilize all those expensive aroma compounds in this application. Extracts should generally be avoided, due to their low solubility in cold, water-based extraction systems. Most extracts commonly require heat to reduce their viscosity, so they will not flow or solubilize in the fermentor at the same level as they would in the whirlpool. Constructed flavors can be employed in a mid-fermentation addition; however, they are generally constructed to be a last-minute addition and their strength is that they can be tailored to provide whichever aromas may be lacking. The uncertain fate of aroma compounds from constructed flavors used during an active fermentation may yield variable and disappointing results.

Post-fermentation Dry Hemp

Post-fermentation dry hemping follows the same overall principles of post-fermentation dry hopping, or, as most of us would simply call it, dry hopping. This addition can happen at any time from when fermentation has just completed to all the way through packaging. In fact, original dry-hopping procedures were developed by British brewers adding whole cone hops to their casks, which bolstered the beer's hop flavor and carbonated the beer through the "freshening power of the hop" (Brown and Morris 1893, 93). Since that time, brewers have taken to adding their hops to the fermentation vessel and allowing the beer to extract the hop aromas prior to packaging. The timing of this addition, coming after fermentation has ended, allows the brewer to extract the highest concentration of aroma compounds into the beer and control the contact time hops stay on the beer to manipulate aroma intensity. I would describe traditionally dry-hopped beers as extracting hops in their most natural state, expressing how we would smell and taste them outside of beer.

Likewise, post-fermentation dry hemping—we'll just call it dry hemping from here on out—presents a great opportunity for brewers to showcase the myriad hemp aromas as we would experience them in nature. Best practices for dry hemping begin with ensuring that fermentation is complete, your apparent extract is stable, and your vicinal diketones (diacetyl) have been reduced below their flavor threshold. At this point, you have two options: either dry hemp at your fermentation temperature, or cool your beer and then dry hemp. From a yeast health perspective, the second option is optimal; your yeast will begin flocculating out of suspension when cooled, so you can either harvest the yeast for another fermentation or simply get it off the beer so it cannot autolyze and add unwanted flavors during maturation and storage. If you maintain fermentation temperatures after dry hemping, you run the risk of triggering a small secondary fermentation and forming off-flavors. Regardless of your finishing method, you are ready to dry hemp.

The extraction efficiency of the beer increases relative to the concentration of ethanol. Like aroma compounds in hops, all hemp aroma constituents favor ethanol as a solvent over water. While terpenoids like linalool may extract better in water-based systems relative to monoterpenes like myrcene, both compounds extract very well in ethanol. Thus, you should pay attention to all the aroma constituents of your cannabis product, because they all are going to go into solution. Looking back to chapter 8 on terpene biochemistry, we saw that terpenes like myrcene, α- and β-pinene, limonene, and β-caryophyllene make up roughly 50–90 percent of the total essential oil in hemp (Mediavilla and Steinmann 1997, 82).

There are several other variables at play that can influence the extraction of hemp aroma compounds during post-fermentation additions. As previously discussed, there is the presence of yeast. Obviously, if fermentation restarts, actively fermenting yeast may affect the balance of aroma compounds through biotransformation. If yeast is still in suspension during dry hemping, it can adsorb monoterpenes and then flocculate them out of solution. A yeast's ability to bind terpenes is roughly correlated to its flocculation rate: the denser and harder the yeast flocculates, the more terpenes it will remove. Non-flocculating yeasts tend to be poorer removers of monoterpenes.

An important variable for influencing hemp aroma extraction is the time and temperature of the extraction. The higher the temperature when dry hemping, the more overall aroma will go into solution. Having aroma intensity is great, but if you are extracting undesirable compounds at the same or greater rate as that of your desired compounds, the quality of your beer's aroma will suffer. Remember, terpenes like myrcene and caryophyllene are highly aroma active and can play a role in masking other desirable compounds. If you favor only efficient extraction, you will more than likely over-extract these terpenes and leave yourself with a one-dimensional, overly piney and grassy beer. You may need to experiment with the balance of time and temperature to help you extract the right ratio of compounds that lead to a complex and delicious aroma. Temperatures lower than 50°F (10°C) will slow down the relative rate of extraction of mono- and sesquiterpenes while still extracting terpenoids like nerol (citrus, rose), linalool, and geraniol at the same rate as a warmer extraction. Eventually, if you extract the hemp product for a long enough time at a lower temperature, the terpenes will fully go into solution, but there is a window of time where you can favor extraction of fruity terpenoids over woody/grassy terpenes.

Finally, as afficionados of hoppy beers know, excess oxygen can play a significant role in a beer's aroma, especially throughout its shelf life. Obviously, excessive amounts of oxygen will lead to stale beer, but even a slightly elevated level of oxygen during dry hemping has the potential to influence hemp aroma even when a beer is fresh. As we've discussed previously, there are many constituents of hemp, such as thiols and thioesters, that are highly aroma-active but exist in low concentrations. These sulfur-based constituents can contribute significantly to the overall impression of a beer, but they are delicate and are some of the easiest compounds to oxidize. A study of 4-methyl-4-mercapto-2-pentanone (4MMP, often described as black currant aroma) in hops found that 4MMP readily transferred from hops to beer and was a significant driver of consumers liking a beer, yet its concentration diminished greatly after warm shelf storage, leading the authors to theorize that oxidative reactions were driving the diminished concentration of 4MMP throughout the beer's shelf life (Reglitz et al. 2018, 99). Interestingly, another study of oxidative effects on hop aroma showed that excess oxygen lowered the sensory perception of tropical, citrus, and hoppy characteristics in a beer, but did not chemically alter the monoterpenes in solution (Barnette and Shellhammer 2019, 179). Hop aroma is complex and its balance can be disrupted by many different factors. There is no reason to suppose hemp aroma is any different. Oxidative compounds like papery and stale aromas can mask your hard-earned hemp aromas and diminish the overall quality of your dry-hemped beer.

Keeping these variables under consideration—presence of yeast, temperature and time of extraction, and control of oxygen ingress during dry hemping—we can now talk about the mechanics of dry hemping. The two big factors are dosage and whether the dry hemping is static or dynamic. Hemp dosage in a beer correlates to the solubility of hemp aroma compounds in that beer; however, since there is only so much alcohol present in which to extract the essential oils from hemp, the relationship between solubility does not follow linearly with the dosage. Stated differently, there is a point at which your beer will become saturated with hemp aroma compounds and adding more hemp product to the beer will not increase its analytical or sensory aroma intensity. The lower your dosage at a given time, the greater the solubility of what you dose, and there is some evidence from hops research that shows if you add smaller doses of hops at multiple addition points you can solubilize more aroma than if you added the same quantity in one (Hauser et al. 2019, 258). The other factor is whether you use static or dynamic dry hemping. Static dry hemping is straightforward: you add your hemp material to your beer and you let extraction happen undisturbed. Generally, static extractions are less efficient, but they are much easier to do and rarely pose the risks of overextraction or oxygen ingress that dynamic extractions do. Dynamic extractions are typically conducted by adding the hemp material and then circulating the beer through a racking arm or standpipe in the fermentor with the aid of a pump. The agitation from the pump along with the moving beer helps extract hemp aromatics faster and more homogenously. The risks to this practice include overextracting your hemp material by circulating too long—often leading to bitter, astringent flavors—and the potential to pick up oxygen through your pump or gaskets while circulating, leading to flavor instability.

The only product not suited for dry hemping is hemp extract, for the same reasons hemp extract is inappropriate for mid-fermentation dry hemping: it is poorly soluble in cold extractions. All of the other products will work well in this process, so long as you have a clear understanding of the aroma constituents of each product and have a good plan as to how you want to go about manipulating those constituents through dry hemping conditions. Hemp flower and hemp pellets will behave in a similar manner as whole hops and hop pellets do during dry hopping, processes that should be familiar to brewers. Hemp essential oils pack a significant concentration of aroma compounds into a small volume and are best used at low concentrations to boost your hemp aroma intensity. It is best to add essential oil to the fermentor to give it some time to homogenize in the tank and to give the beer another chance to mix during transfer to the bright tank. Constructed flavors also are best in dry hemping as the dosage can be fine-tuned; and, if you are unsatisfied with the aroma of your hemp beer, you can work with your vendor to boost certain aromatics you would like.

RECIPES WITH HEMP AROMAS

NEW ENGLAND JUICE BOMB

For 5 US gallons (18.9 L)

Original gravity: 1.072 (17.5°P)
Finishing gravity: 1.016 (4°P)
IBU: 31
Color: 4.5 SRM (8.9 EBC)
Brewhouse efficiency: 75%
Boil time: 70 minutes

This recipe employs many of the modern processing techniques makers of New England-style IPAs use to coax out fruity and juicy aromas from both hemp and hops. Pick any fruity smelling hemp flower you like and experiment with different strains until you find one that gives you the character you are looking for. This recipe relies mostly on additions in the whirlpool and during fermentation to drive yeast biotransformation of hemp and hop terpenoids. As a rule of thumb, try to stay away from the heavy diesel or skunky smelling flowers, as well as flowers that smell overly herbal or spicy.

GRAIN BILL

8.15 lb. (3.7 kg) Pilsner malt
2.64 lb. (1.2 kg) flaked oats
2.2 lb. (1 kg) malted wheat
0.88 lb. (0.4 kg) dextrin malt

HOPS/HEMP

Whirlpool additions

2.58 oz. (73 g) Mosaic (13% AA) @ flame out – whirlpool/steep 15–20 min.
2.50 oz. (71 g) Citra (14% AA) @ flame out – whirlpool/steep 15–20 min.
1.76 oz. (50 g) "Cherry Wine" hemp flower @ flame out
- whirlpool/steep 15–20 min.

Mid-fermentation additions

3.52 oz. (100 g) Citra (14% AA) @ dry hop – ~70% attenuated
0.71 oz. (20 g) "Cherry Wine" hemp flower @ dry hop – ~70% attenuated

Post-fermentation additions

2.12 oz. (60 g) Strata, pellet (12% AA) @ dry hop – maturation vessel
0.18 oz. (5 g) "Cherry Wine" hemp flower (__% AA) @ dry hop – maturation vessel

YEAST

White Labs WLP066 London Fog

DIRECTIONS

1. Mash in at 158°F (70°C) and hold mash for 30 minutes.
2. Runoff and sparge mash with 172°F (78°C) water until you collect 5.68 gal. (21.5 L) in your kettle.
3. Boil for 70 minutes. Depending on your water chemistry, add calcium chloride at the beginning of your boil to target 175–200 ppm calcium chloride.
4. Just prior to flame out, grind up your "Cherry Wine" (or preferred strain) hemp flower to a coarse powder. This will increase the extraction efficiency of the hemp's essential oils.
5. At flame out, add the "Cherry Wine," Mosaic, and Citra whirlpool additions and steep for 15–20 minutes.
6. Cool your wort to 61°F (16°C) and pitch WLP066 yeast. Ferment at 68°F (20°C).
7. When your beer is roughly 70% attenuated, add the Citra and (coarsely ground) "Cherry Wine" hemp flower mid-fermentation additions.
8. Continue fermenting until the gravity is stable over 4–6 days.
9. Cool your beer to 50°F (10°C) and hold for 2 days. Rack the beer off the yeast, hops, and hemp into a clean, carbon dioxide-flushed maturation vessel.
10. Add the Strata pellets and (coarsely ground) "Cherry Wine" hemp flower to the maturation vessel and hold for 2–4 days.
11. Rack the beer off the hops, hemp, and settled yeast; package the beer.

FEELIN' DANKE WEST COAST IPA

For 5 US gallons (18.9 L)

Original gravity: 1.070 (17°P)
Finishing gravity: 1.010 (2.5°P)
IBU: 70
Color: 8.4 SRM (16.5 EBC)
Brewhouse efficiency: 75%
Boil time: 60 minutes

This beer takes the classic West Coast IPA and dials the dank aromas up to 70 (out of 10). Look to use hemp flower that has higher levels of diesel and skunk aromas to play with the old-school West Coast hops.

GRAIN BILL

9.92 lb. (4.5 kg) pale malt
2.65 lb. (1.2 kg) melanoidin malt
0.66 lb. (0.3 kg) caramel 40°L malt

HOPS/HEMP

0.42 oz. (12 g) CTZ (16% AA) @ 60 min.

Whirlpool additions

3.3 oz. (94 g) Simcoe (12% AA) @ flame out – whirlpool/steep 15–20 min.
2.15 oz. (61 g) Summit (16% AA) @ flame out – whirlpool/steep 15–20 min.
2.36 oz. (67 g) CTZ (16% AA) @ flame out – whirlpool/steep 15–20 min.

Dry hop/hemp additions

0.71 oz. (20 g) diesel- and skunk-forward hemp flower, such as AC/DC
2.12 oz. (60 g) Simcoe (12% AA) @ dry hop
1.41 oz. (40 g) CTZ (16% AA) @ dry hop
2.12 oz. (60 g) Centennial (10.5% AA) @ dry hop

YEAST

White Labs WLP001 California Ale Yeast

DIRECTIONS

1. Mash in at 144°F (62°C) and rest for 60 minutes.
2. After the first rest, raise the mash temperature to 169°F (76°C).
3. Run-off and sparge with 173°F (78°C) water until you collect 5.68 gal. (21.5 L) in the kettle.
4. Bring wort to a boil and add CTZ bittering addition. Boil for 60 minutes.

5. At flame out, add whirlpool additions and steep for 15–20 minutes.
6. Cool wort to 59°F (15°C) and pitch WLP001.
7. Ferment at 68°F (20°C) until gravity is stable over 4 days.
8. Cool the green beer to 50°F (10°C), then add the dry hop/coarsely ground hemp additions. Hold for 2–5 days.
9. Rack the beer into a clean, carbon dioxide-purged fermentor or keg. Hold for another 1–2 days to let the yeast settle out, then package the beer.

THIS PILSNER CAME TO PARTY

For 5 US gallons (19.0 L)

Original gravity: 1.046 (11.5°P)
Finishing gravity: 1.006 (1.5°P)
IBU: 30.5
Color: 2.5 SRM (4.9 EBC)
Brewhouse efficiency: 80%
Boil time: 75 minutes

Who says that IPAs get to have all the fun? Pilsners, especially highly hopped Pilsners, make an excellent complement to many hemp strains. Look for hemp strains that smell spicy, woody, earthy, and floral, which will complement the classic German noble hops in this recipe.

GRAIN BILL

8.6 lb. (3.9 kg) Pilsner malt

HOPS/HEMP

0.42 oz. (12 g) German Magnum (14% AA) @ 75 min.
5.29 oz. (150 g) Saaz (3% AA) @ flame out – whirlpool/steep 10 min.
1.16 oz. (33 g) floral hemp flower such as "Lifter" or "Wife" @ flame out – whirlpool/steep 10 min.

YEAST

White Labs WLP800 Pilsner Lager Yeast

DIRECTIONS

1. Mash in at 144°F (62°C) and rest for 75 minutes.
2. Run off and sparge with 169°F (76°C) water until you collect 5.68 gal. (21.5 L) in the kettle.
3. Bring wort to a boil. Add the bittering hops and boil for 75 minutes.
4. Prior to flame out, make sure to grind the hemp flower into a coarse powder.
5. At flame out, add aroma hops and hemp additions to the whirlpool; steep for 10 minutes.
6. Cool the wort to 46°F (8°C) and pitch WLP800.
7. Ferment at 50°F (10°C) until the gravity is stable over 7 days.
8. Cool green beer to 32°F (0°C) and hold for 4 days to flocculate the yeast.
9. Rack the beer off the yeast into a carbon dioxide-purged secondary fermentor. Lager for 4–8 weeks, then package the beer.

THE HERB PANTAGRULION: HEMP GRUIT

For 5 US gallons (18.9 L)

Original gravity: 1.065 (16°P)
Finishing gravity: 1.014 (3.5°P)
IBU: N/A
Color: 7.1 SRM (14 EBC)
Brewhouse efficiency: 75%
Boil time: 90 minutes

This recipe is inspired by the French satirist François Rabelais, who frequently wrote about hemp in defiance of—and mocking—the Catholic Church's banning of cannabis during the Inquisition. This recipe is a mash up of historical beer styles, borrowing from flavors of many ancient farmhouse ales, namely, sahti and gruit. The juniper character of sahti plays nicely with hemp aroma, which I've then bolstered with herbs and spices that could have been found in the old gruits of Europe. I threw in a bit of an old-meets-new twist—the Hornindal kveik that brings aromas of citrus and mango to the beer.

GRAIN BILL
8.6 lb. (3.9 kg) Pilsner malt
3.09 lb. (1.4 kg) Munich malt
0.88 lb. (0.4 kg) hemp hearts

HEMP/SPICES
Bittering addition
0.71 oz. (20 g) fresh dandelion greens @ 30 min.

Whirlpool additions
0.46 oz. (13 g) spicy or herbal smelling hemp @ flame out/whirlpool
0.35 oz. (10 g) juniper berry @ flame out/whirlpool
0.35 oz. (10 g) coriander seed @ flame out/whirlpool
0.18 oz. (5 g) dried lavender @ flame out/whirlpool

Dry hemp addition
0.35 oz. (10 g) fruity smelling hemp flower like "Cherry Wine" @ dry hemp, end of fermentation

YEAST
Omega Yeast OYL-091 Hornindal Kveik

DIRECTIONS

1. Mash in at 140°F (60°C) and rest for 25 minutes.
2. Raise mash temperature to 158°F (70°C) and rest for 30 minutes.
3. Run off and sparge mash with 173°F (78°C) water until you collect 5.8 gal. (22 L) of wort.
4. Boil wort for 90 minutes.
5. At 30 minutes left in the boil, wash your fresh dandelion greens, roughly chop, and then add to the boil.
6. At flame out, grind hemp flower into a coarse powder and add to the whirlpool, along with the coriander seed, juniper berry, and lavender.
7. Cool wort to 68°F (20°C) and pitch OYL-091.
8. Ferment at 86°F (30°C) until the gravity is stable over 5 days.
9. Cool beer to 50°F (10°C), then add dry hemp addition. Extract for 2 days, then rack beer off yeast and hemp into a clean, carbon dioxide-purged maturation vessel.
10. Age for 5–7 days, then package.

11

QUALITY ASSURANCE AND LEGAL COMPLIANCE (UNITED STATES)

GENERAL CANNABIS QUALITY CONSIDERATIONS

An old brewing fable begins with a brewmaster who shows up at a malthouse for a surprise inspection. The brewmaster does not bring a pen and notepad, but rather a slice of bread. At the outset of the inspection, the brewmaster tells the maltster that if the malt house is truly clean and worthy of beer, then the brewmaster could swab any surface of the malthouse with the slice of bread and ask the maltster to take a bite. While this is draconian, the sentiment of this fable provides a good framework for how everyone in the beer supply chain should approach hygiene in their manufacturing environment.

Hemp quality begins on the farm. This starts with recognizing both the limits and opportunities for farming operation improvements. Anyone who has spent time on a farm knows that they are naturally dirty places. Food contaminants abound, from literal dirt and mold to bugs and pests to manure and fertilizers. Such is the nature of, well, nature. However, food quality from farming operations has dramatically improved over the last fifty years due to dedicated efforts by farmers and regulators to eliminate food pathogens and improve food processing conditions.

Hemp presents a unique challenge to brewers. Caught up in the "gold rush" that is the current CBD business, many farmers are planting an unfamiliar crop they've never processed before, and even amateur farmers with little to no experience in farming, let alone food processing, are planting hemp. In such a new industry and with ever-changing regulations, quality will fluctuate from farm to farm. It is important for brewers who work with farms to articulate their expectations for handling and quality practices so they can have confidence in what they will eventually serve their customers. By no means should brewers behave like the fabled brewmaster and show up with a slice of bread, but they can pass on their practical experience working with hop and barley farmers to share best practices.

There are several great organizations that work on raw material quality for the brewing industry in the US, including the American Malting Barley Association (AMBA), Hop Research Council (HRC), Hop Quality Group (HQG), Hop Growers of America (HGA), and Brewers Association (BA). I strongly encourage you to visit those websites and review their publications to better familiarize yourself with their respective areas of focus. Collectively, these groups have centuries of brewing experience and I urge you to join and participate with these groups to help them continue their work in producing high-quality brewing materials.

Additionally, several international certification bodies provide helpful information on criteria for quality, as well as certifying farmers and food processors that meet those criteria. The three main organizations are the USDA's Good Agricultural Practices (GAP),[1] the International Organization for Standardization (ISO), and the American Herbal Products Association, whose aim is to globally improve and standardize best practices for many industries related to growing and processing products for the food and beverage industries. While it is optimal for farmers and hemp processers to have certifications from these organizations, this expectation is often unrealistic due to the high costs associated with the certification process, with many small farms simply not having enough revenue to pursue it. However, even in the absence of such certifications, there are still many practical considerations farmers and producers can take on board to dramatically improve the safety and quality of processing hemp.

---

1 Farmers can request a GAP audit from the USDA at their website, https://www.ams.usda.gov/services/auditing/fruits.

It's best practice to visit all your raw material suppliers from time to time, and in the case of hemp I suggest that you visit farms both during the growing season and harvest to better understand how the farm operates. Below is a short list of what to look for and discuss during your visit. Two important points to remember: you are their guest and you are there to strengthen your partnership. Berating a farmer about their practices is counterproductive to building your relationship. Your partners don't walk into your brewery and point out everything they think you are doing wrong; extend the same courtesy to them.

What to look for at a hemp production farm:

- Is the farm generally tidy? Are tools and equipment stored in an intended place or left haphazardly?
- Does the farm take special consideration for using cleanable, food-safe materials with their equipment? For example, if the farm uses belts and conveyors, are they lubricated with food-safe lubricants?
- Are the processing facilities clean? Do they have adequate lighting and ventilation? Is there a dedicated processing area?
- How is the hemp processed and dried? Does the farm try to limit the ingress of birds and pests that could eat or defecate on the drying hemp?
- How is the dried hemp stored? Does the facility limit the hemp's exposure to light and oxygen? Is it stored in a way that pests can't get at it? Is it stored in a way that mold does not develop?
- How many people touch the hemp? Does the farm have plans in place for food-safe handling practices, such as wearing hairnets and gloves?
- Does the equipment look safe? Do the workers have appropriate personal protective equipment? Does the farm take their workers' health and well-being into account?
- What efforts does the farm take to produce hemp in an environmentally sustainable way?

LABORATORY TESTING FOR HEMP QUALITY COMPLIANCE

As you work with a hemp farm to determine the ideal harvest date and drying conditions, you should also learn about the farm's external laboratory testing protocols. The tests outlined in the sections that follow are required by law to guarantee food safety and legal compliance. Which tests are needed for *Cannabis* may vary from state to state, so it is important to familiarize yourself with all local, state, and federal testing guidelines. These should be considered as the baseline for quality and legal compliance, but there are several other specialized tests—such as essential oil analyses, mycotoxin and general microorganism contamination screening, and residual solvent testing (for extracts)—that can also be done depending on the concerns of your organization. Typically, the farm is responsible for testing its hemp for cannabinoid and contaminant compliance; however, you and your organization should follow up with both the farm and the laboratory to ensure they are following best practices.

The cannabis plant has a remarkable ability to take up nutrients from the soil, some of which may be detrimental to human health. I've heard anecdotes from farmers who have been surprised to learn that their hemp crop pulled out pesticides and heavy metals from their soil that they had no idea were even there. It's important, especially when working with farmers that are new to hemp, that they apply a comprehensive battery of tests to their hemp and provide you with the results so that you do not inadvertently add something toxic or illegal to your beer. I discussed this issue in more detail in chapter 4 (pp. 105–6), but I will repeat here the points I made there:

Any cannabis grown for phytoremediative purposes should not be used for human consumption. If you are sourcing any cannabis material for use in products destined for human consumption, be sure you understand your supply chain and how producers should operate and test for contaminants in harvested plant matter.

Fortunately, many independent testing laboratories work specifically with the cannabis industry to provide high-quality testing that assures you and your consumers purchase a safe product. Unfortunately, due to the uncertainty of hemp and marijuana's legal status over the last ten years, no universal regulator oversees compliance in these laboratories, so there is some variability in the precision of results. Most laboratories work with both hemp and marijuana—since marijuana is federally prohibited, these labs cannot be audited by the Food and Drug Administration (FDA). Typically, state departments of agriculture step into that role; however, in the absence of federal oversight, there are no universal protocols outlining best practices and independent auditing for laboratories across states. In this situation, testing facilities are left to establish their own internal best practices, which may differ from laboratory to laboratory.

Amidst this uncertainty, it is essential for brewers to exercise due diligence by reviewing their testing facilities to make sure the laboratories have proper protocols in place. Laboratories that invest in quality programs are easy to spot and frequently advertise their quality compliance programs. Quick ways to spot these quality-focused providers include accreditation from an outside certification body, such as the ISO; a documented Good Manufacturing Practice (GMP) and Good Laboratory Practice (GLP) program; a documented process and method validation program to assure test accuracy and repeatability; and an easy to understand chain of custody program to track testing. If you find a testing facility is not familiar with any of these programs, find another laboratory that is.

TESTING FOR CANNABINOIDS AND TERPENES

Testing the level of cannabinoids in your hemp flower is essential to ensure the product you are using is federally compliant and is not above the legal threshold for marijuana, which is defined as cannabis plant material containing 0.3 percent or more delta-9 tetrahydrocannabinol (THC, or Δ^9-THC) by weight. Farmers in the US test their crop for cannabinoid potency while it is in the field to determine the harvest date and to file their compliance paperwork with the relevant state's department of agriculture. Often, farmers will provide the results of these tests to their customers to show compliance. The problem is that this test often happens weeks before the hemp is ultimately harvested and, even though it shows legally compliant hemp, THC levels may have risen sufficiently in the interim to move the material out of compliance. Thus, it is important to test your hemp again after it is harvested and dried to make sure that you are not inadvertently purchasing something that is legally defined as marijuana. This test for dry weight potency is also important for your Alcohol and Tobacco Tax and Trade Bureau (TTB) and FDA paperwork when you file for label approval.

Laboratories frequently use high-performance liquid chromatography (HPLC) or ultra-high-performance liquid chromatography (UPLC) for cannabinoid quantification; however, other technologies such as quantitative nuclear magnetic resonance (qNMR) can also be used successfully. A typical report will contain the primary cannabinoids—CBG, CBD, THC, and CBC—in both their acidic and neutral forms. Some tests may also include noteworthy psychoactive cannabinoids such as CBN and Δ^8-THC (both THC artifacts), as well as exotic variants like THCV and CBDV (see chap. 5 for a discussion of all these cannabinoids). The report should list the total concentrations for each individual cannabinoid as well as the ratio of cannabinoids to one another.

Many laboratories also quantify terpene content, either as its own test or in conjunction with cannabinoid sampling. There are no legal testing requirements

CERTIFICATE OF ANALYSIS

Prepared for:
New Company
123 Main Street
Anywhere, ST USA 00001

Beverage Additive, 900mg

Batch ID or Lot Number: **00001**	Test: **Potency**	Reported: **12Mar2022**	USDA License: N/A
Matrix: Concentrate	Test ID: T9999999	Started: 11Mar2022	Sampler ID: N/A
	Method(s): TM14 (HPLC-DAD)	Received: 11Mar2022	Status: N/A

Cannabinoids	**LOD** (%)	**LOQ** (%)	**Result** (%)	**Result** (mg/g)	**Notes**
Cannabichromene (CBC)	0.026	0.090	0.230	2.30	
Cannabichromenic Acid (CBCA)	0.024	0.082	0.030	0.30	
Cannabidiol (CBD)	0.078	0.257	2.860	28.60	
Cannabidiolic Acid (CBDA)	0.080	0.264	0.230	2.30	
Cannabidivarin (CBDV)	0.018	0.061	ND	ND	
Cannabidivarinic Acid (CBDVA)	0.033	0.110	ND	ND	
Cannabigerol (CBG)	0.015	0.051	0.060	0.60	
Cannabigerolic Acid (CBGA)	0.061	0.214	ND	ND	
Cannabinol (CBN)	0.019	0.067	ND	ND	
Cannabinolic Acid (CBNA)	0.042	0.146	ND	ND	
Delta 8-Tetrahydrocannabinol (Delta 8-THC)	0.073	0.255	ND	ND	
Delta 9-Tetrahydrocannabinol (Delta 9-THC)	0.066	0.231	0.110	1.10	
Delta 9-Tetrahydrocannabinolic Acid (THCA-A)	0.059	0.205	ND	ND	
Tetrahydrocannabivarin (THCV)	0.013	0.046	ND	ND	
Tetrahydrocannabivarinic Acid (THCVA)	0.052	0.181	ND	ND	
Total Cannabinoids			**3.520**	**35.20**	
Total Potential THC**			0.110	1.10	
Total Potential CBD**			3.062	30.62	

Final Approval

Signature	Analyst Name 18Mar2022 12:42:00 PM MDT	*Signature*	QA Name 18Mar2022 12:53:00 PM MDT
PREPARED BY / DATE		APPROVED BY / DATE	

Definitions
% = % (w/w) = Percent (weight of analyte / weight of product). ND = None Detected (defined by dynamic range of the method).
Total Potential Delta 9-THC or CBD is calculated to take into account the loss of a carboxyl group during decarboxylation step, using the following formulas: Total Potential Delta 9-THC = Delta 9-THC + (Delta 9-THCa *(0.877)) and Total CBD = CBD + (CBDa *(0.877)).

Testing results are based solely upon the sample submitted to Botanacor Laboratories, LLC, in the condition it was received. Botanacor Laboratories, LLC warrants that all analytical work is conducted professionally in accordance with all applicable standard laboratory practices using validated methods. Data was generated using an unbroken chain of comparison to NIST traceable Reference Standards and Certified Reference Materials. This report may not be reproduced, except in full, without the written approval of Botanacor Laboratories, LLC. ISO/ IEC 17025:2005 Accredited A2LA

ilac-MRA A2LA ACCREDITED ASTM

Figure 11.1. A mock certificate of analysis (COA) as they commonly appear. Not all COAs will provide this much information. Image supplied courtesy of Botanacor Laboratories™.

for terpenes; however, it is a helpful analysis to define the key aroma analytes in your hemp and to devise strategies for extracting those aromas in your beer (chap. 10). Terpene analyses vary from laboratory to laboratory depending on their equipment and methodological sophistication. A laboratory should provide a list of terpenes it will quantify; if you see a terpene or terpenoid not listed in the standard test battery, you may have to request that compound or find another vendor that will quantify it. The common method used to measure terpenes utilizes gas chromatography (GC), typically combined with mass spectroscopy (GC-MS); however, GC can be combined with other detection methods, such as flame ionization detection (GC-FID), vacuum ultraviolet spectroscopy (GC-VUV), and olfactometry (GC-O). These additional methods may aid in identifying and quantifying certain hemp-derived aroma compounds.

TESTING FOR CONTAMINANTS

Contaminant testing is essential to guarantee that your hemp flower is a food-safe ingredient for your beer. Such tests are common in the food and beverage industry and mirror testing protocols set by the FDA. These tests seek to identify contaminants accumulated during the growth, harvest, and storage of hemp, which can include heavy metals, residual pesticides, and harmful bacteria and fungi that can produce deadly mycotoxins.

While useful as an environmental phytoremediator, the downside of cannabis's ability to hyperaccumulate heavy metals and minerals from soil means it might not be suitable for producing products that are fit for human consumption. Years of growing other crops—with the attendant additions of herbicides, pesticides, and fungicides—can cause a build up of heavy metals in the soil that goes undetected. Hemp commonly accumulates most heavy metals during the first planting. The planted hemp can pull toxic compounds into the plant, providing an unwelcome surprise to the farmer at harvest, especially if the farmer is unaware of the presence of these compounds. Testing for heavy metals like lead, mercury, arsenic, and cadmium screens potential contaminated hemp material before it enters the food and beverage supply chain. Every hemp and marijuana crop should have a heavy metals test with each harvested lot, or as directed by the relevant state's department of agriculture. The test report should include the metals tested, the limits of quantification and detection for the method, and the results. If your hemp lot does not have a heavy metals test associated with it, send a sample to an accredited laboratory to confirm it is safe before using it.

Hemp may also accumulate agricultural pesticides during growth, even if they are not directly applied. As a sticky, resinous, wind-pollinated crop, hemp is particularly susceptible to windblown pesticides. Many common crops receive pesticides via aerial crop dusting that may blow into adjacent fields. While these chemical applications may be approved for use on the targeted crops, the compounds may stray onto a hemp crop, potentially exceeding thresholds such that, if humans ingest these pesticides, they may be carcinogenic and mutagenic. Currently, no pesticides are approved for hemp in the US, although that is likely to change in the coming years. Hemp farmers should keep detailed records of any direct pesticide applications, including time and dose; further, they should test their harvest for residual pesticides, which will look for most USDA-approved pesticides along with their detection and quantification limits.

Hemp may also contain bacterial or fungal pathogens that can produce potentially fatal toxins. Farmers often add manure to their fields as a nitrogen supplement for their soils; if that manure contains harmful enterobacteria (e.g., salmonella or *E. coli*), those pathogens can transfer to and survive on the plant matter. If the farm has a good quality program in place and handles its hemp in a food-safe manner, the risk of pathogen transmission is low.

Pathogen detection involves swabbing and plating a sample on a series of microbiological growth media to try to grow any pathogens present. Should bacteria or fungi grow on the media, the sample is analyzed using the polymerase chain reaction (PCR) to determine the exact species to inform further testing. Most tests will return some amount of bacterial or fungal contamination, represented as total colony forming units (CFUs); however, most of these microorganisms so identified are benign to humans. Indeed, some brewers see varying CFUs from their dry-hopped beers, attributable to the dry hop additions. Regardless, while most of these CFUs represent microorganisms that are benign to beer and humans, it is important to make sure of this. Some specific fungal species, such as aspergilli and penicillia, can produce additional mycotoxins that can be carcinogenic, neurotoxic, hepatotoxic (liver damaging), and more. These species can be detected through microbiological plating and confirmed through PCR. Additional testing to quantify any mycotoxins specific to the fungal species occurs after initial detection and identification.

The presence of excessive mold CFUs in a hemp sample is a leading indicator for poor drying or storage conditions on the farm and should be grounds for rejecting a hemp lot. Lots should have associative pathogen testing prior to storage; however, depending on storage conditions, additional testing should be considered prior to shipping. Understanding the order of operations for how your vendor processes, tests, and stores hemp will help you determine any additional testing required prior to receiving a lot.

CERTIFICATE OF ANALYSIS

Prepared for:
New Company
123 Main Street
Anywhere, ST USA 00001

Beverage Additive, 900mg

Batch ID or Lot Number:	Test:	Reported:	USDA License:
00002	**Pesticides**	**18Mar2022**	NA
Matrix:	Test ID:	Started:	Sampler ID:
Concentrate	T999990	17Mar2022	NA
	Method(s):	Received:	Status:
	TM17 (LC-QQ LC MS/MS)	14Mar2022	NA

Pesticides	**Dynamic Range** (ppb)	**Result** (ppb)
Abamectin	285 - 2836	ND
Acephate	22 - 2768	ND
Acetamiprid	38 - 2784	ND
Azoxystrobin	36 - 2744	ND
Bifenazate	36 - 2731	ND
Boscalid	39 - 2849	ND
Carbaryl	34 - 2710	ND
Carbofuran	35 - 2707	ND
Chlorantraniliprole	44 - 2727	ND
Chlorpyrifos	43 - 2796	ND
Clofentezine	280 - 2758	ND
Diazinon	281 - 2752	ND
Dichlorvos	265 - 2775	ND
Dimethoate	37 - 2798	ND
E-Fenpyroximate	272 - 2833	ND
Etofenprox	37 - 2778	ND
Etoxazole	298 - 2775	ND
Fenoxycarb	37 - 2723	ND
Fipronil	27 - 2755	ND
Flonicamid	53 - 2744	ND
Fludioxonil	320 - 2719	ND
Hexythiazox	34 - 2817	ND
Imazalil	292 - 2686	ND
Imidacloprid	36 - 2813	ND
Kresoxim-methyl	49 - 2805	ND

	Dynamic Range (ppb)	**Result** (ppb)
Malathion	282 - 2728	ND
Metalaxyl	36 - 2724	ND
Methiocarb	36 - 2754	ND
Methomyl	40 - 2777	ND
MGK 264 1	162 - 1661	ND
MGK 264 2	107 - 1140	ND
Myclobutanil	44 - 2733	ND
Naled	36 - 2746	ND
Oxamyl	39 - 2784	ND
Paclobutrazol	42 - 2674	ND
Permethrin	299 - 2786	ND
Phosmet	36 - 2722	ND
Prophos	276 - 2727	ND
Propoxur	37 - 2719	ND
Pyridaben	294 - 2747	ND
Spinosad A	32 - 2241	ND
Spinosad D	51 - 503	ND
Spiromesifen	274 - 2844	ND
Spirotetramat	267 - 2730	ND
Spiroxamine 1	16 - 1176	ND
Spiroxamine 2	20 - 1553	ND
Tebuconazole	270 - 2725	ND
Thiacloprid	37 - 2779	ND
Thiamethoxam	37 - 2821	ND
Trifloxystrobin	36 - 2747	ND

Final Approval

Signature
Analyst Name
18Mar2022
12:42:00 PM MDT
PREPARED BY / DATE

Signature
QA Name
18Mar2022
12:53:00 PM MDT
APPROVED BY / DATE

Definitions
ND = None Detected (defined by dynamic range of the method)
Dynamic Range = Limit of Quantitation (LOQ) through Upper Limit of Method Range
ppb = Parts Per Billion

Testing results are based solely upon the sample submitted to Botanacor Laboratories, LLC, in the condition it was received. Botanacor Laboratories, LLC warrants that all analytical work is conducted professionally in accordance with all applicable standard laboratory practices using validated methods. Data was generated using an unbroken chain of comparison to NIST traceable Reference Standards and Certified Reference Materials. This report may not be reproduced, except in full, without the written approval of Botanacor Laboratories, LLC. ISO/IEC 17025:2005 Accredited A2LA.

ilac-MRA ACCREDITED ASTM Certified Test Laboratory

Botanacor Laboratories, LLC. | © All Rights Reserved | 1301 S Jason St Unit K, Denver, CO 80223 | 888.800.8223 | www.botanacor.com

Figure 11.2. A mock certificate of analysis (COA) for contaminants. Image supplied courtesy of Botanacor Laboratories™.

CERTIFICATE OF ANALYSIS

Prepared for:
New Company
123 Main Street
Anytown, ST USA 00001

Beverage Additive, 900mg

Batch ID or Lot Number:	Test:	Reported:	USDA License:
00002	**Heavy Metals**	**18Mar2022**	NA
Matrix:	Test ID:	Started:	Sampler ID:
Concentrate	T999990	15Mar2022	NA
	Method(s):	Received:	Status:
	TM19 (ICP-MS): Heavy Metals	14Mar2022	NA

Heavy Metals	**Dynamic Range** (ppm)	**Result** (ppm)	**Notes**
Arsenic	0.04 - 4.32	ND	
Cadmium	0.04 - 4.42	ND	
Mercury	0.04 - 4.41	ND	
Lead	0.04 - 4.17	ND	

Final Approval

Signature	Analyst Name 17Mar2022 11:29:00 AM MDT	*Signature*	QA Name 17Mar2022 11:36:00 AM MDT
PREPARED BY / DATE		APPROVED BY / DATE	

Definitions
ND = None Detected (defined by dynamic range of the method)
Dynamic Range = Limit of Quantitation (LOQ) through Upper Limit of Method Range

Figure 11.3. A mock certificate of analysis (COA) for heavy metals.
Image supplied courtesy of Botanacor Laboratories. Image supplied courtesy of Botanacor Laboratories™.

HEMPSEED QUALITY ASSURANCE AND COMPLIANCE

Hempseed Quality Considerations and Testing

Due to the high content of unsaturated fatty acids in its oil, hempseed will go rancid if not properly stored. Most considerations relating to hempseed flavor quality therefore involve proper storage and limiting staling in stored hempseed.

Most issues with hempseed quality can be resolved through good communication with your vendor and tight control of your supply chain inventory. Good communication and documentation from your vendor should illuminate any quality issues that may arise, and you can take proactive steps to periodically review the documentation with your vendor. To maintain legal compliance in the US, hempseed vendors must sterilize their hempseeds so that they cannot sprout and must verify the hempseeds do not contain residual THC greater than 0.3 percent by weight after harvest and cleaning. Vendors should provide supporting documentation with the date, time, and method of sterilization and cleaning, plus THC potency testing results before shipping you a hemp seed lot. This paperwork must be included when submitting a formula for approval with the TTB. From a quality perspective, your hempseed vendor should also provide documentation stating the harvest date, storage conditions, expected shelf life, and shipping protocols for each lot.

Harvested hempseed requires extra care and consideration when storing, so it is best to put the onus on an organization that manages those conditions daily ("push the problem on the vendor"). Brewers should practice just-in-time inventory management with hempseed by ordering just enough for production demand. Ordering in bulk and storing hempseed in a silo like malted barley runs the risk of staling and rancidity.

If you do decide to purchase hempseed in bulk, ensure proper storage using cold storage, limiting oxygen ingress, and limiting light exposure. At the very least, hempseed should be stored at refrigeration temperatures, ideally below 50°F (10°C). At those temperatures, hempseed should keep for a minimum of 10–12 months. For extended shelf life and storage, hempseed may be frozen, which can extend its shelf life for another 12 months. Hempseed should also be stored in light-impermeable bins with a firm, fitting lid. This sufficiently limits oxygen ingress and light exposure and keeps out any pests. Extra efforts, such as flushing the atmosphere of the bin with nitrogen or carbon dioxide and sealing the container, will provide optimal storage stability. Following these best practices should give your hempseed a shelf life of 24 months.

Prior to milling, conduct a visual inspection of your hempseed to look for mold, bugs, and debris or other detritus. Additionally, taste the hempseed for any staling or rancid flavors. If the seeds taste papery, rancid, cheesy, or just generally off, discard them and order another lot. The compounds responsible not only taste bad but will also introduce other staling precursor peroxide compounds that will detrimentally affect the shelf life of your beer.

If you have the capability to do further analysis of your hempseed, refer to AOAC International's Official Methods of Analysis to better get an idea of the diversity of analytical methods available for quantifying hempseed constituents (https://www.aoac.org). Additionally, there are many third-party laboratories that provide nutritional analyses of food products, including calorie count, carbohydrate analysis, oil and fat content, mineral content, moisture, and others. You can look online to find local or regional laboratories that do nutritional testing in your area. Your malt vendor may be able to analyze your hempseed for parameters commonly measured in malt, including moisture content, total extract, total protein, total oils, ash, mineral content, and color (the last one especially useful if seeds are toasted or roasted). Politely inquire with your malt vendor if they have these capabilities or if they have recommendations for third-party laboratories that might be able to help.

Hemp Seed Legal Considerations

The Alcohol and Tobacco Tax and Trade Bureau (TTB), an organization with which most brewers are probably familiar, is a regulatory agency within the US Department of the Treasury that is responsible for the taxation and regulation of alcoholic products. As part of that regulatory oversight, the TTB must approve all alcoholic beverage formulations before they can be legally produced, labeled, and sold. To ease the regulatory burden, the TTB regularly publishes a list of preapproved "exempt" ingredients allowed in beer[2] so that regulators do not have to pore over every ingredient of every formulation for the tens of thousands of beer recipes that cross their desk. While hempseed is an ingredient allowed by the TTB to be in beer, it is not one of the ingredients on the exempt list. Therefore, brewers must file a formula for any hempseed beer with the TTB to demonstrate compliance with current federal restrictions, which mostly concern the prohibition of psychoactive substances such as THC in alcoholic beverages. To see a general overview of the TTB's hemp policy, visit https://www.ttb.gov/formulation/hemp-policy.

---

[2] The TTBs most recent list of exempt ingredients can be found on their website. Please check to see if this has been updated. https://www.ttb.gov/images/pdfs/rulings/ttb-ruling-2015-1-attachment-1.pdf.

When submitting your hempseed beer formula to the TTB, you will need the residual THC testing data and documentation of seed sterilization from your vendor. Read your vendor's test results to make sure that it contains the name and contact information of the laboratory, a detailed description of the THC detection method, the result of the THC quantification test, and the minimum detection limit of THC associated with the testing method. If any of this information is missing from your formula it will be rejected and you will have to file supplemental paperwork to meet the TTB's requirements. Additionally, brewers may be required to send a sample to a TTB-approved laboratory to verify no cannabinoids are present in the final beer.

Once the TTB approves your formula, you can submit your certificate of label approval. The TTB should give some guidance on accurately labeling your hempseed inclusion and suggest variations. Be careful during the labeling process to be diligent in how you speak about your beer's hempseed inclusion and that you do not overtly or subtly suggest that marijuana is in the beer or that a consumer will get a psychoactive effect. The TTB reviews each label on a case-by-case basis—imagery of marijuana paraphernalia or overt marijuana references may be grounds for rejection.

HEMP AROMA QUALITY ASSURANCE AND COMPLIANCE

Hemp Aroma Quality Considerations

Due to the obvious parallels, a high-quality hemp compliance program should closely resemble a quality assurance program for hops. Several organizations provide many excellent resources for brewers implementing a hop supply chain quality program, including the BA, HQG, HRC, and HGA (see p. 218). I would encourage you to make use of these resources when building a hemp quality program and to start thinking about which criteria are most important for you and your business.

At this point, I'm assuming you have taken the time to vet several farms for the quality of their hemp and how they run their operations. If your chosen farm is local, it is best to visit several times during the growing season to see how the crop is developing. As your hemp reaches maturity, there may be opportunities to see how the crop's aroma is developing and to start making notes as to when it may be mature. The hop industry employs a small army of specialists to conduct this work based on years of industry experience; unfortunately, at this point in time, there is no such parallel in the hemp industry. As a brewer looking to get specific aromas from your hemp, you may have to do more work to evaluate the hemp lots while they are still in the field and see if your desired aroma characteristics are at their proper levels. Of course, you should pair this work with a qualified laboratory, who can quantify your desired aroma compounds and track their development as the hemp matures in the field.

Every farmer that I have worked with says that terpene development in the cannabis plant waxes and wanes throughout its lifecycle. Farmers report that terpenes tend to develop early and are highest in concentration between four to six weeks before the plant reaches full maturity. This makes intuitive sense considering that we know terpenes are biochemical intermediates in cannabinoid biosynthesis; so, as more cannabinoids develop, the concentration of terpenes will drop. Harvesting hemp early to take advantage of the apex concentration of terpenes may be suitable for certain brewers depending on their aroma criteria. The downside to harvesting hemp early is that the terpene profile generally lacks many of the oxidation-driven compounds, such as terpenoids, that drive more complex aromas (see chapters 8 and 10). If you are unsure of what you want, you can ask your farmer to provide some samples for you from the early, middle, and last harvests to ascertain which timing best suits your flavor development goals.

Familiarity with harvest timings will also allow you to take advantage of sensory techniques employed by brewers for hop selection. Grinding or rubbing fresh hemp in a similar manner to rubbing whole leaf hops can give you an excellent idea of the aromatic quality of a hemp sample. You will want the hemp to be representative of what you will use, so you may need to work with your farmer to set processing guidelines ahead of time to get a representative sample. If the dried sample contains excessive amounts of fan leaves or sugar leaves, the hemp may possess an overly grassy smell, which may also mask the aroma compounds from the glandular trichomes. Observations like these will help give an early indication if your farmer is going to process your hemp in a way that deviates from your expectations.

On top of specifying how you would like excessive foliage trimmed from your flower, you should have a conversation about your expectations for drying, trimming, and packaging of your finished hemp flower. Hop drying is a heavily engineered, precision process; you will not find that precision in the cannabis industry. You should have a conversation about how dry you would like your hemp to be, especially to prevent mold formation. The importance of this will differ for brewers operating in humid environments like North Carolina than in more arid environments like the Yakima Valley or Colorado. If your farm has the capability to measure moisture content, you should target about 10 percent moisture, plus or minus a few percent. As good practice, especially if you live in a humid climate, have your vendor package your hemp with a material that does not allow moisture ingress, like vacuum-sealed mylar. Next, you should specify your trimming parameters, setting how much leaf and stem content is acceptable. You are paying for plant material containing aroma compounds; you do not want to spend good money on excessive material that does not contribute desirable aromas to your beer. Hops suppliers call this specification the leaf and stem content on the certificate of analysis. Finally, discuss how your hemp will be packaged, shipped, and stored. Like hempseed, your hemp flower should be stored in a cold, moisture-controlled environment, away from direct light and ideally in an oxygen-deficient environment. This will provide the longest shelf life and should allow you to store your hemp for up to two years without a significant drop in aroma quality. It may be asking too much of your farm to invest in packaging infrastructure suited for a brewer's needs, so you may have to compromise. If the hemp can be packed in workable quantities in vacuum-sealed bags, that will improve quality. If the hemp can be packed in nitrogen-flushed, oxygen-barrier packaging such as mylar, even better. At the very least, your hemp should stay in a cool environment and in a container impervious to light.

When brewing with hemp, think about the quality considerations you would apply when brewing IPAs. If your brewery has sophisticated techniques for quantifying terpenes and volatile aromatics this will be a benefit. For smaller operations, focus your efforts on basic beer quality and sensory analysis. Maintaining low levels of dissolved oxygen in beer, maintaining healthy yeast, and practicing good hygiene around the brewery goes a long way in hemp-flavored beers. Taste your beer at regular intervals throughout its shelf life to determine aromatic instability and identify problem areas. To set true-to-brand targets, you should do some basic attribute definition and scaling work at the outset. These targets will help you identify potential problem areas in your process and catch flawed beer before it goes out the door. Software such as DraughtLab® can help you create a sensory program and design tests to further improve your attribute recognition.

Hemp Aroma Legal Considerations

The primary legal consideration for registering hemp-flavored beers with the TTB is gaining approval for a novel, non-exempt ingredient. The approval process involves properly filing a flavor ingredient data sheet (FID) that must be evaluated by both the TTB and FDA. For this reason, using constructed hemp flavors has an advantage over other products because flavor houses have large regulatory teams that file FIDs for you. For the basis of this discussion, we will set aside constructed flavors from flavor houses that have not been derived from cannabis because those flavors have a good track record of FID approval. If you do explore natural flavors, make sure to ask if the flavor has a TTB-approved FID before evaluating the sample so you don't waste time and energy.

Include the FID with the broader beer formula so TTB has all of the information it needs. The TTB uses the FID to verify that you will not add any ingredient in excess of FDA recommended limits, and that the TTB can give appropriate labeling guidance. The ingredient should be available to be independently tested by TTB laboratories to assure accuracy and compliance.

This is the moment where tracking and documenting your quality compliance tests pays off. These tests are intended to give the TTB and FDA a precise understanding of what is in your hemp sample, at least regarding anything that may pose a health risk (e.g., contaminants) and bioactive substances such as cannabinoids. Like hempseed, the government cares the most about the risk of adding THC into your beer. Tests should demonstrate that your hemp aroma product contains less than 0.3 percent THC by weight and thus qualifies as being industrial hemp. Additionally, use a dosage calculator similar to the calculator outlined in the tincture section (p. 153) to help estimate how much theoretical THC could be dosed based on the inclusion rate. Since you know that THC is mostly insoluble in beer without the aid of an emulsifier, you should also test a sample of hemp-dosed beer to bolster your case that no THC survived into the final beer. In all these tests, you will want to include a detailed summary of the detection method used and its limits of detection and quantification; the test should be validated for the type of sample analyzed. Contaminant testing should show that your hemp is safe as a food product and that it carries little risk of contaminating your beer with residual heavy metals, pesticides, or solvents.

If you get your hemp product approved, it should be designated as generally recognized as safe (GRAS) for use in food and beverages. The approval process will likely take a few rounds of back and forth with the TTB and FDA before they issue final approval. With an approved formula, you can then file your certificate of label approval and begin producing your beer. The same considerations for labeling compliance for hempseed beers also apply to hemp-flavored beer; avoid overt or subtle drug references and suggestions that the beer will elicit a psychoactive effect.

CBD COMPLIANCE

CBD Quality Considerations

To commercially produce beers and beverages with consistent levels of CBD from package to package, it is imperative to use concentrated, emulsified CBD nanoparticles (a.k.a. CBD concentrates or "water-soluble CBD"). Homebrewers are encouraged to use these products as well but can also find good success with the other techniques outlined in chapter 7.

Working with concentrated, emulsified cannabidiol (CBD) products involves all the quality considerations outlined so far in this chapter, except that you must test the concentrate prior to dosing, as well as test the finished product. The process of concentrating cannabinoids will result in any contaminant present in the extracted biomass also being concentrated, so quality approaches should seek to identify potential problems before extraction and confirm successful results post-extraction. Initially, you should focus quality audits on the extractor's quality program, testing compliance, and food safety processes; once you are sufficiently satisfied with those parameters, you can work with your CBD vendor to audit their upstream partners.

Producers of water-soluble CBD emulsions are a different type of organization than hemp farmers and should be held to different standards. For all intents and purposes, the production of water-soluble CBD emulsions is a blend of food and pharmaceutical manufacturing: your producer's processes should reflect that. High-quality manufacturers will have well-documented quality compliance programs, strict food safety standards, a dedicated environmental health and safety program, a litany of quality certifications (e.g., GMP, Hazard Analysis at Critical Control Points [HACCP], and ISO 9000 or 22000), and documentation showing they have passed safety and health inspections from local, state, and federal inspection agencies. You may find organizations without this level of sophistication; my advice is stick to organizations that can clearly articulate their food safety and quality programs and demonstrate that quality is their primary focus.

As you search for a vendor, ask for samples of products and also any associated certificates of analysis (COA). The COAs should include all ingredients in the formulation, shelf-life and storage recommendations, physical characteristics of the sample, and test results, including tests for cannabinoid potency, microbiological test results, heavy metals test results, residual solvent test results, and residual pesticide test results. For indirect contaminants not associated with the process of manufacturing the concentrate—such as heavy metals, solvents, and pesticides—the COA may just give a threshold that the manufacturer guarantees the sample to be under. This information should be auditable upon request.

Once satisfied that your CBD vendor has the quality and testing chops to be worthy of your beer, you can then test the organoleptic and physical stability of the emulsified water-soluble product. I recommend starting with flavor threshold testing.[3] Today, most beverages range in inclusion rates from five to fifty milligrams of CBD per serving; however, inclusion rates may vary in the future based on research and consumer demand. For a flavor threshold test, select a target CBD concentration, then make a series of samples at up to five times the intended concentration to see how each one tastes. CBD concentrates may have associative flavors with them, depending on factors like the carrier oil, the emulsifier type, and residual essential oil from the CBD extract. You should be able to ascertain at which concentration you notice flavors deleterious to your base beer, allowing you to develop flavor mitigation strategies if need be. The common refrain is that water-soluble CBD products tend to have some bitterness and astringency; the better you can test and quantify these attributes, the easier it will be to plan to mitigate them. Brewers may not have issues with these qualities, as most beers tend to have some degree of bitterness and astringency.

Once you understand the organoleptic qualities of the CBD formulation, conduct a series of physical stability trials with the concentrate. Because pH, the presence of various metal cations, oxidation, light exposure, and heat exposure can all drive instability in a CBD emulsion, it is best to conduct these trials in conjunction with your CBD vendor to see if you get reproducible results. Spend some time thinking about the life of your beverage after it leaves the loving embrace of your facility. How will it be transported? How will it be stored? How long will sit on the shelf at a store? When a customer ultimately purchases it, how will that person enjoy it? As discussed in chapter 7, putting CBD-dosed beer in kegs presents additional challenges. Given a little time sitting at cold temperatures, the CBD emulsion will layer out and, theoretically, 100% of all the CBD originally in the keg could end up in a single pint. Following the recommendations in chapter 7 for periodically rousing the keg will help alleviate this problem but cannot guarantee consistency from pint to pint. Therefore, I suggest using package formats where you can control the content of cannabinoids in each serving, such as bottles and cans, or look to point of dispense dosing strategies to guarantee pint to pint CBD concentrations.

---

[3] A reminder: it is good form to notify tasters that the samples contain CBD, a pharmacologically active molecule.

Can Liners and Cannabinoid Emulsions

A notable phenomenon that has been found to occur in can-packaged cannabis beverages is where the hydrophobic core of the emulsion particles cause some of the emulsion to bind to the can liner, removing a portion of the cannabinoid from the beverage during its shelf life. Many can manufacturers are working on new can liner compositions that resolve this issue, but no solution is currently in place. Unfortunately, nothing can be done at this moment, but it is good to be aware. The best prevention is sourcing high-quality, stable emulsions and adopting manufacturing practices that limit emulsion instability. You should test the degree to which can liners will bind cannabinoid emulsions in conjunction with your can vendor to best understand the impact on the potency and shelf life of your product.

Beverage distribution can be rough on product shelf life, so it helps to design tests that will best simulate the worst-case scenario. I like to think of the "height of summer" scenario, where your distributor's HVAC system is struggling to keep up with 100°F temperatures outside, your CBD drink gets loaded onto an unrefrigerated truck and driven around for seven hours in the heat of the day, unloaded onto a hot loading dock, merchandised and stocked on a warm shelf where it sits for months, then purchased and thrown in a cooler at a beach party. We all pray that scenario exists only in our nightmares, but we know it happens. Designing a test for that scenario could look something like this:

1. Select your intended package type (bottle or can) for the shelf-life study.
2. Dose a large enough quantity of beer to be studied—kegs or small bright tanks work—and thoroughly mix to homogenize the sample.
3. Use a counter-pressure filler, or packaging line, to fill individual sample bottles.
4. Partition off enough replicate samples into at least three lots: warm stored (at least 70°F, or 21°C), combined warm and cold stored, and cold stored (40°F, or 4°C, or below).
5. Pull samples at one-month intervals and visually inspect for emulsion stability. If you have the capability, check the cannabinoid potency to see if there are changes.
6. Re-do the test with any new formulation that differs with respect to pH or changes in processing, such as diatomaceous earth filtration (which causes iron pickup in beer).

While conducting your shelf-life testing, it is worthwhile validating the calculated cannabinoid concentration. This will both verify the tests conducted by your vendor and help identify any variability in your process that could cause you to fall out of compliance range. Conducting multiple tests ensures that the quantification of your cannabinoids is stable and reproducible. If you have HPLC or UPLC technology in your brewery, you can conduct this kind of quality testing in-house. Waters Corporation recently published a UPLC method that you can find on their website (https://www.waters.com) that will help develop this capability. If you do not have this capability in-house, send samples both to your CBD vendor as well as a third-party laboratory. If you're a real stickler, you can also ask the third-party laboratory to run an independent sample for heavy metals, pesticides, and residual solvents to validate your vendor's COA.

At this point, you have tested the organoleptic properties of your CBD concentrate, tested the physical stability of your CBD-dosed beer in a variety of temperature conditions, and verified the cannabinoid concentration both in your bulk sample and in your dosing test. You can now move into full-scale production. After reading chapter 7, you should have a good plan for how you will dose and homogenize your CBD concentrate into a batch. As you package, follow a repeatable sampling protocol from your packaging run to verify you are getting stable, reproducible results. A good place to start is to calculate how long your packaging run is likely to take and plan to sample at even time intervals during the run. At the very least, take samples at the beginning, middle, and end of your packaging runs and send the samples out for cannabinoid testing to make sure you are not seeing premature settling in your packaging feed tank.

CBD Legal Considerations

In the US, at the time of writing this in 2022, CBD is not approved by the FDA to put into any food or beverage. Before I go any further, I want to state clearly and unequivocally that if you are a commercial manufacturer looking to produce CBD beverages, you should not do so until the FDA issues a clear ruling on adding CBD to food and drinks. Bummer, I know. You may be asking yourself, especially if you live in a pro-cannabis state, why can I purchase products containing CBD at my grocery/liquor/convenience store? Why do I see ads to purchase CBD products online? The answer is that these products are legal for you to purchase and there are no legal repercussions to you, the consumer, for possessing or consuming them.

I realize the previous paragraph is probably confusing, especially considering you've just read through most of a book telling you strategies on how to do a thing I'm now telling you not to do. I share your confusion and will attempt to articulate why everything is the way it is. I'll start with the non-confusing part. If you're a homebrewer who wishes to add CBD to your beer, go nuts. Purchasing and possessing CBD isn't illegal. Use the information provided in this book to go make some delicious CBD-infused concoctions—share them with your friends and enjoy all the wonderful things cannabis has to offer to you, a personal consumer.

The confusing part applies to manufacturers and retailers, all of whom are taking legal risks by marketing and selling CBD-infused products. Stated again, it is currently illegal under federal law to produce, market, and distribute CBD-infused products. There are many opinions on how severe the risks are to these organizations; I will not speculate about which opinion is correct. The crux of this issue pertains to the role the US federal government and state governments play in regulating food and beverages. Keep in mind, their role is to protect consumer health and safety and make sure that manufacturers adhere to clear rules and standards when producing their products. The FDA, in its last press release,[4] continues to state that it is studying the safety of CBD as a food, beverage, or dietary supplement. While the FDA outlined several concerns about the safety of CBD, it is less clear what information it needs to issue a final ruling, or for how long it would like to continue to study CBD. This state of affairs has been ongoing since the passage of the 2014 Agriculture Act, which initially allowed for the production of hemp and hemp products for research purposes. Now, almost a decade later, and in the absence of clear guidance from the FDA, many companies have decided that they will make CBD products and hope that the federal government does not penalize them for doing so. There are currently plenty of CBD products on the shelf—time will tell if they are recognized as a legal product or if they will be removed permanently.

There is hope that this legal ambiguity may soon be at an end. The US House of Representatives recently introduced bill H.R. 6134, the CBD Product Safety and Standardization Act of 2021. If enacted, it would establish federal standards and require the FDA to regulate CBD in food and beverages. The FDA falls under the executive branch of the federal government, so the agency's regulatory focus can change with each presidential administration. However, Congress can pass legislation to force agencies like the FDA to pass permanent

4 "What You Need to Know (And What We're Working to Find Out) About Products Containing Cannabis or Cannabis-derived Compounds, Including CBD," U.S. Food and Drug Administration, last updated March 5, 2020, https://www.fda.gov/consumers/consumer-updates/what-you-need-know-and-what-were-working-find-out-about-products-containing-cannabis-or-cannabis.

rulings that comply with enacted federal legislation. If H.R. 6134 becomes law, the FDA will have to issue regulations specifying the maximum amount of hemp-derived CBD per serving, labeling and packaging requirements for CBD products, and conditions for intended use.[5] With these regulations in place, producers and retailers will finally be able to comply with the law and provide the CBD products consumers want.

---

5 Food and Drug Law at Keller and Heckman, "House Introduces Bill to Regulate CBD in Food," *National Law Review,* December 8, 2021, https://www.natlawreview.com/article/house-introduces-bill-to-regulate-cbd-food.

EPILOGUE

Throughout this book, it has been my primary objective to give useful information about the status of cannabis as it is today in the US, and how that pertains to brewers interested in using cannabis in their products, at least to the extent that this is legally possible. Now, I'd like to attempt to read the tea leaves and talk about the future of cannabis and what that means for the beer and beverage industries. I feel a bit like early craft beer pioneers, who recognized the vast opportunities that flavorful beer presented in the US, but who could not fully imagine how those opportunities would manifest. Considering the explosive growth of craft beer in America, the crazy things that have come to pass, and the knock-on effects in other markets, I try to imagine what is to become of the cannabis industry, acknowledging that much of it is beyond my comprehension in these early days.

I genuinely believe that craft beer has a role to play in the cannabis industry, as a thought leader and thoughtful listener. Cannabis needs highly educated, quality-focused professionals who can responsibly promote an intoxicant and minimize its societal downsides. Craft brewing needed energetic idealists who sought to disrupt the norms of the culture and industry they found themselves in. The similarities between the two businesses are too many to ignore; both should look to deepen their connections with one another for mutual health and longevity.

Cannabis acceptance and legalization is on the march. Many respected economists forecast both hemp and marijuana to grow at double-digit compounded annual growth rates for the next decade. Aside from electric cars and renewable energy, the 2020s may very well be defined by the explosive growth of cannabis. Estimates of cannabis's global economic impact sit between $50 billion and $100 billion annually, much of which exists in an unregulated, untaxed, and unaccountable black market.

Furthermore, marijuana continues to be used by governments and law enforcement as a cudgel to harm communities of color. This has led to the enrichment of violent drug cartels and needless suffering. Voters, sensing a better way, are creating a significant backlash against drug policies that have been shown to be ineffective at lowering use rates and to actively destroy communities. Voters are demanding change. According to a 2019 Pew Research poll, 67 percent of Americans think that marijuana use should be legal, a higher proportion than ever before. Young people are leading this surge of acceptance and will likely be the drivers of the new cannabis economy.

To build this new economy, the focus will have to remain on innovating high-margin products that allow smaller niche markets to emerge. At the outset of writing this book, I spoke to a Canadian entrepreneur, Stephen Christiansen, about his patented hemp decortication technology, HempTrain™ Advanced Processing System. Few people have innovated on hemp decortication technology since the US Civil War, for the many reasons described in this book. One insight Stephen left with me was the need for early innovators to focus on high-margin goods to help create additional viable cannabis opportunities. The HempTrain system is designed to optimally process the three major hemp outputs: CBD from the inflorescences and leaves, specialty fiber products from the stalk, and hurds from the stalk core. This allows farmers to maximize their crop yield and make sure nothing goes to waste. Of course, the CBD delivers the best return on the crop, so the HempTrain design focuses on high CBD yields and capturing waste streams that other CBD processors miss. Focusing on the highest-margin cannabis commodities combined with efficiently utilizing the whole plant will be crucial for the long-term success of cannabis.

Ultimately, we will have to address the disparate and unsustainable ways *Cannabis sativa* is grown. Indoor marijuana operations have huge environmental footprints due to their use of high-wattage grow-lights and their sizable waste streams. Too many outdoor marijuana operations are illegally run in national forest land where they clear-cut the forests and pollute the rivers to grow their crop. Far too many hemp operations care only about CBD production, which generates large amounts of processing waste and delivers few ecological benefits. If cannabis is to truly live up to its reputation as an environmentally friendly crop, we must efficiently grow and harvest it.

I interviewed some local hemp breeders and we eventually took the discussion to what an ideal hemp or marijuana plant looks like. Right now, cannabis plants grown for cannabinoids tend to create poor-quality fiber and fiber plants tend to make lower levels of cannabinoids. To the breeders, the ideal cannabis plant would look like a cattail. It could be grown at the same density as fiber hemp varieties, carrying some of the ecological benefits of dense planting. The bushy, cannabinoid rich inflorescence could be mechanically harvested, then the stalk could be cut down for processing into myriad products from fiber and hurds. In this idyllic world, these crops could be grown on any soil type in most temperate climates of the world. We would no longer have to destroy treasured forests in search of a quick buck, nor would we have to go through herculean efforts to grow genetic freaks in an industrial warehouse using otherwise unheard-of amounts of chemical fertilizers.

Cannabis needs predictability. If predictability is going to come for the cannabis industry in America, it must first come from the US government. The current patchwork regulatory scheme rewards those who wish to test the limits of the law and punishes those who want to take—forgive the pun—a sober-minded approach to innovating these products. Currently, small companies take outsized legal risks to innovate and tend to do so in ways that open them to multiple levels of legal and even criminal liability. Many companies operate as if no regulators are paying attention and no one will punish them if they are caught. Look no further than an FDA study examining the total CBD content of labeled products: one in five of the products studied contained less than 80 percent of the CBD indicated and nearly two in five contained *more* than 120 percent of the CBD indicated (Hahn 2020, 6). Would you or your customers tolerate a beer that had that level of variability? The patchwork regulatory framework surrounding the cannabis trade allows for these kinds of shenanigans to occur and, unfortunately, due to funding issues, enforcement is spotty at best. The only businesses that tend to meet labeling specifications are those with the financial ability to do so, generally large businesses or businesses with large financial backers. Do not confuse this as a treatise advocating for the need for big business to enter the cannabis market; this is about quality and safety, and our current regulatory scheme does a lackluster job of incentivizing both. Can you imagine if you walked into a bar, ordered a drink, and the drink could leave you somewhere between completely sober and twice the legal driving limit?

To be fair, the marijuana market has done a much better job of regulating itself than the hemp market. This is attributable to states taking an outsized role in regulation and enforcement, knowing that if they allow too much, the federal government will come in and shut their program down. With billions flowing into local economies and state coffers, state governments cannot afford to be flippant with regulation. Even then, states like California have not been able to control the black and grey marijuana markets, which has allowed unregulated marijuana products to flourish. Tragically, this led to deadly consequences, where marijuana vaporizer cartridges were sold with an unapproved carrier, vitamin E acetate. Vitamin E acetate is an FDA-approved, GRAS compound for use in food, beverages, and topical applications; however, when it is vaporized and inhaled into the lungs, it can cause lung damage colloquially described as "popcorn lung." Without uniform regulatory guidelines and a common regulatory body that provides both oversight and enforcement, these kinds of incidences will continue to happen.

Beyond a predictable regulatory program, cannabis products—whether containing THC or CBD—need to provide a predictable effect. Early legal cannabis markets brought such wild variance to the user experience that it is hard to build commonality among consumers. If consumers continue to get wildly different experiences from similar products with similar dosages, how can you expect them to take it consistently? Many of the early influencers in the marijuana business tended to be some of the heaviest users and as a result probably gave a lot of bad advice to first-time customers. I've had the experience in Colorado where I was told a 50-milligram (mg) edible should give me a pleasant, gentle high. Luckily, I knew that even a 10 mg edible is a bit much for me and politely declined the budtender's suggestion. Many new consumers do not have enough information to know what dose will be most effective for them. Infamously, Maureen Dowd wrote in the *New York Times* ("Don't Harsh Our Mellow, Dude," June 3, 2014) about her bad experiences eating a marijuana brownie in a hotel room in Denver—these types of stories will scare away potential new consumers. Without predictability in the cannabis consumer market, fewer opportunities present themselves for it to be accepted in mainstream culture.

There are published guidelines for alcohol serving sizes and ways to measure intoxication; at this point no such analogous advice exists in the marijuana trade. Without these health guidelines, we cannot recommend how to avoid driving intoxicated, let alone allow marijuana products to be served in a bar or restaurant. Some municipalities are exploring opening tasting rooms where consumers can consume marijuana in a legal and safe fashion; however, few tasting rooms have opened to date. My guess is these tasting rooms haven't taken off because there are no agreed upon serving standards, let alone reproducible ways for detecting intoxicated driving. In the absence of these standards, municipalities will not shoulder the risk of approving businesses that may directly endanger public health. Should common serving and intoxication standards develop, it will go a long way toward normalizing marijuana and will allow it to enter more public spaces. Reflecting on the dramatic increase of socially acceptable places to serve beer over the last thirty years, it's not a stretch to imagine that cannabis could follow a similar arc. Making cannabis predictable—both legally and experientially—will open a new world of opportunities.

Cannabis needs a diverse customer base. There are already signs of the beer and cannabis industries coming together, from resource sharing between the hops and hemp industries to brewers producing cannabis-infused products. The hop industry already has a lot of the technical know-how and processing infrastructure to produce high-quality hemp. Hop farms require significant investment capital to start and maintain and have highly specialized equipment. Hop farmers spend millions of dollars on such equipment, which typically sits idle for ten to eleven months of the year because there is no other crop for which it can be redeployed. Hemp could change all that and is proven to be a suitable second crop to process on hop farming equipment. Hemp tends to be harvested in the northern hemisphere between September to the middle of October; generally, this is after the last hops have been picked. A demand for hemp could extend the window of operation for hop farmers' equipment, increasing profits for their operations. Profitable farms can in turn re-invest in their operations and maintain the levels of quality in their hops that craft brewers need for their beers.

Additionally, the hop industry already has equipment tailor-made to service the brewing industry. Significant opportunities exist for hemp operations that are looking to diversify their customer base to partner with hops brokers and farmers and tap into the brewing business. Likewise, hops brokers can provide high-quality liquid carbon dioxide-extracted concentrates for the CBD business in the short term and possibly the THC business in the long term. Their established processing and storage capabilities can preserve raw cannabis material far better than upstart operations and they have the quality programs in place to satisfy stringent food safety regulations. Global GAP (https://www.globalgap.org) is fast becoming the industry standard for hop growers; this program presents a significant improvement in the standardization of farming and food handling practices among hop growers and should be promoted as the standard for farmers globally. Hop growers do not have to directly become hemp growers for a spillover effect to occur in the hemp industry; rather the brewing supply chain should present Global GAP as its preferred standard for every farm in its supply chain and help encourage farmers to take steps toward meeting those standards.

There are also openings for the hops industry to explore cannabis industry technologies, especially around water-soluble emulsions. Emulsified hop essential oils present a potential for aromatically impactful, lossless, and stable products for brewers. For small brewers especially, hops additions contribute significantly to beer losses in the cellar. Products derived only of hops that deliver true-to-type aromas present a large opportunity for brewers to reduce losses in the cellar while still delivering world-class beers to their customers. Additionally, these products can dramatically lower shipping costs and associated environmental impacts. Emulsions are not a new concept to hops, in fact there were many commercial emulsified hop products available to brewers throughout the 1960s and 1970s, particularly for adding hoppy notes to British-style ales (Laws et al. 1978, 69). These historical emulsified hops oils had many technical challenges and were limited in use. The advancement of new emulsification technologies, combined with the demand for efficient, loss-reducing hop products, presents a new opportunity for the brewing community to explore.

Hopefully, by now, this book has got you thinking about some of the opportunities cannabis presents for your brewery. The primary focus obviously surrounds the use of hemp to make interesting, differentiated beers that continue to push the boundaries of what is possible in craft beer. In my personal experience, I've met countless drinkers who are interested in cannabis aromas and seek different ways to pair beer with their cannabis. This presents a chance for us to build upon those experiences. We've all seen the parabolic growth curve of IPA, mostly fueled by the introduction of new flavors and styles that pair well with hops. Cannabis is a new ingredient that behaves the same way as hops but delivers flavors that hops cannot. In my ideal world, we will find new ways to develop the flavors of both and create a whole new category of beer. I've brewed many iterations of this beer and I can assure you it is delicious and tremendously exciting.

Additionally, hemp can push the brewing community to think about its environmental footprint and promote new ways of doing business. Products like hempcrete could allow a business not only to run their operations more efficiently, but also be part of the solution in drawing down atmospheric carbon dioxide emissions. As hemp becomes a larger, more diversified crop, hemp-based products like packaging materials, biodegradable plastics, and durable work wear become more cost effective. Considering hemp's efficiency and environmental benefits, the sooner we can develop this industry, the sooner we can reduce our environmental footprint. Make no mistake, the vitality of the beer business is intertwined with society's ability to mitigate climate change. Already barley and hop farms globally have lost crops due to catastrophic weather events, which will only become more frequent with climate change. If we allow this to continue, we will eventually see crops become uneconomically expensive due to losses and risks on the farm. Obviously, hemp is not the only solution to fix climate change, but it should be considered as part of the systemic overhaul necessary to efficiently grow, process, and produce many of the goods that a modern world requires.

Cannabis presents a disruptive risk to the alcohol industry. Recent studies show that young people consider marijuana to be a safer drug to consume than alcohol and that it poses fewer long-term health risks when used responsibly. Whether that position is true is beside the point. As marijuana becomes more available to the public, the greater the risk of disruptive replacement. New Frontier Data found that 45 percent of people in a studied focus group indicated they were likely to replace some of their drinking with cannabis in the future and 65 percent stated that, if given the choice, they would prefer marijuana over alcohol (De Carcer 2019, 7). This should send shivers down the spine of anyone in the alcohol industry. However, these insights do present opportunities for forward-thinking businesses who are thinking about how they might be able to enter this space. There are already brewers manufacturing cannabis-infused beverages to varying levels of success. Brewers such as Lagunitas, Ska, and Flying Dog have entered the marijuana market with infused beverage products that reflect some of their core offerings. Due to the patchwork legal framework in the marijuana business, these products are manufactured by a third party who possesses a permit to manufacture and sell marijuana-infused products, instead of these breweries producing marijuana-infused beverages on the same equipment as their beers. Additionally, most of these products currently exist as a licensing agreement, where the brewery licenses their brand out to the manufacturer. This keeps a degree of separation between the brewery that must comply with federal regulators and the marijuana manufacturer who can operate on the state level. In a world where marijuana is legal under federal law and regulated accordingly, I would guess many of these barriers will come down and you will see more direct manufacturing of marijuana-infused beverages by breweries.

Brewers are also developing marijuana expertise by investing in Canadian marijuana companies and forming joint research and development programs for manufacturing marijuana-infused products. Large global brewers, such as Constellation Brands (Corona), MillerCoors, Anheuser-Busch, and Molson, have entered joint partnerships with Canadian marijuana manufacturers to test and develop products in the new legal marijuana market in Canada. At this point in time, infused beverages make up a small fraction of total sales in Canada; however, these large brewers remain bullish on the future of these products. Ostensibly, there are several hurdles that these companies need to overcome, like making great tasting beverages that people want to drink and that deliver a predictable effect. Next, they need to spend the effort to convince consumers that marijuana-infused beverages are worth considering. You can't drink a marijuana beverage in a bar (or in public for that matter) so there's less of a social niche that beverages traditionally fill. If marijuana were to be allowed in bars or in dedicated tasting rooms, these barriers may come down quickly. There is a general health consensus that is moving consumers away from smoking; logically, one could conclude that vaping will not be far behind. As traditional forms of marijuana consumption like smoking and vaping move out, it presents many more openings for food and beverage products to move in and supplant the demand for smokable products.

With hemp and CBD, there are obvious opportunities for brewers and beverage manufacturers to exploit. CBD users are now joining a broad group of consumers who are looking for products that promote their health and wellness. New Frontier Data lists the top two reasons consumers take CBD as pain management and relaxation/stress relief. This follows many concurrent consumer trends that show people are looking for healthier alternatives to their current food and beverage choices and are willing to pay a premium for these products. Non-alcoholic beer sales have seen significant growth in the last few years, driven by increasing acceptance of a sober lifestyle. CBD seems to fit neatly within this paradigm as a product advertised to aid in relaxation and promote better sleep, with fewer downsides than taking stronger psychoactive drugs like alcohol and marijuana.

Living in Colorado, I see many beverage products now available, including a new line of CBD seltzers made by Left Hand Brewing Company, named Present. I chatted with Left Hand CEO Eric Wallace and he echoed many of the concerns and opportunities I've articulated here. Currently, Present's largest hurdle is the lack of regulatory clarity from the FDA. This limits the brand's ability to gain further distribution and limits the brewer's opportunities to develop a new beverage category.

The original inspiration for Present, Eric tells me, was to produce a viable alternative product to the hard seltzer craze. Left Hand had little interest in developing a hard seltzer and chose to take a different approach to the category. Eric, along with many at Left Hand, is an avid cyclist and soccer player; he finds that CBD soothes aching muscles and helps him relax. He wanted to expand the company's horizons and develop other beverages that would keep Left Hand drinkers excited and bring new drinkers into the fold. While hard seltzer didn't excite him or his staff, CBD did.

In my opinion, too many of the CBD infused beverages—and THC infused beverages—look and feel like kids' drinks. Many come in sweet, soda-like flavors that have less appeal for an older audience. I think there is an opportunity to reinvent the non-alcoholic category to include adult-oriented flavors that are suitable for happy hour. The recent successes of non-alcoholic spirits like Seedlip show that sober adults want products that reflect their age. These products also allow the sober and the sober-curious to still have access to spaces where alcohol is served without feeling socially ostracized for their sobriety. CBD-infused beverages present a huge opportunity to add to this space by offering adults a relaxing adult beverage without the intoxication or the hangover.

While cannabis may disrupt alcohol's role in society, I doubt it will replace it. Alcohol—but beer especially—has been a global cultural cornerstone for several millennia. Beer connects humans: we drink to celebrate, to socialize, to bring people together. So far, through no fault of its own, cannabis remains disconnected from these community building activities. Societally, we prefer to smoke marijuana in private; that notion is reflected in our laws and our habits. Its roots as a symbol of counter-cultural rebellion continue to this day. Cannabis still does not have the same level of access or acceptance as alcohol. For it to truly grow, someone will have to figure out how to bring the ritualization and community of alcohol to cannabis. I hope this book can contribute to bringing this future to bear.

SUGGESTED READINGS

The following is a partial list of published works and sources that I am highlighting to illustrate many of the themes central to this book.

CANNABIS HISTORY

Abel, Ernest. 1980. *Marijuana: The First Twelve Thousand Years.* New York: Plenum.

Booth, Martin. 2003. *Cannabis: A History.* New York: Picador.

Clarke, Robert, and Mark Merlin. 2013. *Cannabis: Evolution and Ethnobotany.* Berkeley: University of California Press.

Lee, Martin A. 2012. *Smoke Signals: A Social History of Marijuana—Medical, Recreational, and Scientific.* New York: Simon and Schuster Press.

Robinson, Rowan. 1996. *The Great Book of Hemp.* Rochester: Park Street Press.

Russo, Ethan. 2007. "History of Cannabis and Its Preparations in Saga, Science, and Sobriquet." *Chemistry and Biodiversity* 4(8): 1614–1648. https://doi.org/10.1002/cbdv.200790144.

Small, Ernest. 2017. *Cannabis: A Complete Guide.* Boca Raton: CRC Press.

RACIAL AND CRIMINAL JUSTICE

Alexander, Michelle. 2010. *The New Jim Crow: Mass Incarceration in the Age of Colorblindness.* New York: The New Press.

American Civil Liberties Union. 2013. *The War on Marijuana in Black and White.* New York: ACLU Foundation.

Blow, Charles. 2010. "Smoke and horrors." *New York Times*, October 22, 2010.

Blow, Charles. 2011. "Drug bust." *New York Times*, June 10, 2011.

Bouie, Jamelle. 2014. "The case for marijuana reparations." *Slate*, July 28, 2014.

Frederique, Kassandra. 2021. "To truly reimagine safety, we must end the war on drugs." *Washington Post*, March 16, 2021.

Koram, Kojo, ed. 2019. *The War on Drugs and the Global Colour Line.* London: Pluto Press.

Pfaff, John. 2017. *Locked In: The True Causes of Mass Incarceration.* New York: Basic Books.

Provine, Doris. 2007. *Unequal Under Law: Race in the War on Drugs.* Chicago: University of Chicago Press.

TECHNICAL CANNABIS PUBLICATIONS

Abrahamov, Aya, Avraham Abrahamov, and Raphael Mechoulam. 1995. "An Efficient New Cannabinoid Antiemetic in Pediatric Oncology." *Life Sciences* 56(23–24): 2097–2102. https://doi.org/10.1016/0024-3205(95)00194-B.

Andre, C., J.-F. Hausman, and G. Guerriero. 2016. "*Cannabis sativa*: The Plant of the Thousand and One Molecules." *Frontiers in Plant Science* 7:19. https://doi.org/10.3389/fpls.2016.00019.

ElSohly, M. and Slade, D. 2005. "Chemical Constituents of Marijuana: The Complex Mixture of Natural Cannabinoids." *Life Sciences* 78(5): 539–548. https://doi.org/10.1016/j.lfs.2005.09.011.

Hazekamp, Arno, Justin Fischedick, Mónica Llano Díaz, Andrea Lubbe, and Renee Ruhaak. 2010. "Chemistry of Cannabis." In *Development and Modification of Bioactivity*. Vol. 3 of *Comprehensive Natural Products II: Chemistry and Biology*, edited by Hung-Wen (Ben) Liu and Lew Mander, 1033–1084. Leiden: Elsevier. https://doi.org/10.1016/B978-008045382-8.00091-5.

Hillig, K., and P. Mahlberg. 2004. "A Chemotaxonomic Analysis of Cannabinoid Variation in *Cannabis* (Cannabaceae)." *American Journal of Botany* 91(6): 966–975. https://doi.org/10.3732/ajb.91.6.966.

Page, Jonathan E., and Jana Nagel. 2006. "Biosynthesis of Terpenophenolic Metabolites in Hop and Cannabis." In *Integrative Plant Biochemistry*, edited by J. T. Romeo, 179–210. Recent Advances in Phytochemistry, vol. 40. Kidlington, Oxon.: Elsevier. https://doi.org/10.1016/S0079-9920(06)80042-0.

Russo, E., and G.W. Guy. 2006. "A Tale of Two Cannabinoids: The Therapeutic Rationale for Combining Tetrahydrocannabinol and Cannabidiol." *Medical Hypotheses* 66(2): 234–246. https://doi.org/10.1016/j.mehy.2005.08.026.

Russo, E. 2011. "Taming THC: Potential Cannabis Synergy and Phytocannabinoid-Terpenoid Entourage Effects." *British Journal of Pharmacology* 163(7): 1344–1364. https://doi.org/10.1111/j.1476-5381.2011.01238.x.

WHO. 2018. *WHO Expert Committee on Drug Dependence: Critical Review; Cannabis and Cannabis Resin*. Report prepared for 41st Expert Committee on Drug Dependence. Geneva: World Health Organization.

APPENDIX A: HEMP FARM VISIT CHECKLIST

EQUIPMENT AND PROCESSING FACILITIES

- ❑ Is the farm generally tidy?
- ❑ Are tools and equipment stored in an intended place or left haphazardly?
- ❑ Does the farm take special consideration for using cleanable, food-safe materials in their equipment? For example, if the farm uses belts and conveyors, are they lubricated with food-safe lubricants?
- ❑ Are the processing facilities clean?
- ❑ Do the processing facilities have adequate lighting and ventilation?
- ❑ Is there a dedicated processing area?

STORAGE

- ❑ How is the hemp processed and dried?
- ❑ Does the farm try to limit the ingress of birds and pests that could eat or defecate on the drying hemp?
- ❑ How is the dried hemp stored?
- ❑ Does the facility limit the hemp's exposure to light and oxygen?
- ❑ Is the hemp stored in a way that pests can't get at it?
- ❑ Is the hemp stored in a way that mold does not develop?

SAFE FOOD HANDLING PRACTICES AND EMPLOYEE SAFETY

- ❑ How many people touch the hemp?
- ❑ Does the farm have plans in place for food-safe handling practices, such as wearing hairnets and gloves?
- ❑ Does the equipment look safe?
- ❑ Do the workers have appropriate personal protective equipment?
- ❑ Does the farm take their workers' health and well-being into account?

SUSTAINABILITY

- ❑ What efforts does the farm take to produce hemp in an environmentally sustainable way?

BIBLIOGRAPHY

Abel, Ernest L. 1980. *Marihuana: The First Twelve Thousand Years*. New York: Plenum.

Abrahamov, Aya, Avraham Abrahamov, and Raphael Mechoulam. 1995. "An Efficient New Cannabinoid Antiemetic in Pediatric Oncology." *Life Sciences* 56(23–24): 2097–2102. https://doi.org/10.1016/0024-3205(95)00194-b.

ACLU (American Civil Liberties Union). 2013. *The War on Marijuana in Black and White*. Policy Paper. New York: ACLU Foundation.

Amaducci, S., A. Zatta, M. Raffanini, and G. Venturi. 2008. "Characterisation of Hemp (*Cannabis sativa* L.) Roots Under Different Growing Conditions." *Plant and Soil* 313:227–235. https://doi.org/10.1007/s11104-008-9695-0.

Ames, F. 1958. "A Clinical and Metabolic Study of Acute Intoxication with *Cannabis Sativa* and its Role in the Model Psychoses." *Journal of Mental Science* 104(437): 972–999. https://doi.org/10.1192/bjp.104.437.972.

Anderson, E. 1954. *Plants, Life, and Man*. London: Melrose.

Anderson, D. M., B. Hansen, D. I. Rees. 2019. “Association of Marijuana Laws with Teen Marijuana Use: New Estimates From the Youth Risk Behavior Surveys.” *JAMA Pediatrics* 173(9): 879–881. https://doi.org/10.1001/jamapediatrics.2019.1720.

Andre, C. M., J-F. Hausman, and G. Guerriero. 2016. “*Cannabis sativa*: The Plant of the Thousand and One Molecules.” *Frontiers in Plant Science* 7:19. https://doi.org/10.3389/fpls.2016.00019.

Appendino, G., S. Gibbons, A. Giana, A. Pagani, G. Grassi, M. Stavri, E. Smith, and M. M. Rahman. 2008. “Antibacterial Cannabinoids from *Cannabis sativa*: A Structure–Activity Study.” *Journal of Natural Products* 71(8): 1427–1430. https://doi.org/10.1021/np8002673.

Barnette, B. M., and T. H. Shellhammer. 2019. “Evaluating the Impact of Dissolved Oxygen and Aging on Dry-Hopped Aroma Stability in Beer.” *Journal of the American Society of Brewing Chemists* 77(3): 179–187. https://doi.org/10.1080/03610470.2019.1603002.

Benelli, G., R. Pavela, R. Petrelli, L. Cappellacci, G. Santini, D. Fiorini, S. Sut, S. Dall’Acqua, A. Canale, and F. Maggi. 2018. “The Essential Oil from Industrial Hemp (*Cannabis sativa* L.) By-Products as an Effective Tool for Insect Pest Management in Organic Crops.” *Industrial Crops and Products* 122:308–315. https://doi.org/10.1016/j.indcrop.2018.05.032.

Benet, Sula. 1975. “Early Diffusion and Folk Uses of Hemp.” In *Cannabis and Culture,* edited by Vera Rubin, 39–50. World Anthropology. The Hague: De Gruyter Mouton. https://doi.org/10.1515/9783110812060.39.

Bonnet, U., M. Specka, M. Soyka, T. Alberti, S. Bender, T. Grigoleit, L. Hermle, et al. 2020. “Ranking the Harm of Psychoactive Drugs Including Prescription Analgesics to Users and Others—A Perspective of German Addiction Medicine Experts.” *Frontiers in Psychiatry* 11: 592199. https://doi.org/10.3389/fpsyt.2020.592199.

Bonnie, R.J., and C.H. Whitbread. 1970. “The Forbiden Fruit and the Tree of Knowledge: An Inquiry into the Legal History of American Marijuana Prohibition.” *Virginia Law Review* 56(6): 971–1203.

Booth, J. K., and J. Bohlmann. 2019. "Terpenes in *Cannabis sativa* – From Plant Genome to Humans." *Journal of Plant Science* 284:67–72. https://doi.org/10.1016/j.plantsci.2019.03.022.

Booth, J. K., J. E. Page, and J. Bohlmann. 2017. "Terpene Synthases from *Cannabis sativa*." *PLoS ONE* 12(3): e0173911. https://doi.org/10.1371/journal.pone.0173911.

Booth, Martin. 2003. *Cannabis: A History*. New York: Picador.

Borougerdi, Bradley Jahan. 2014. *Cord of Empire, Exotic Intoxicant: Hemp and Culture*. PhD diss., University of Texas, Arlington. http://hdl.handle.net/10106/26673.

Brenneisen, Rudolf. 2007. "Chemistry and Analysis of Phytocannabinoids and Other *Cannabis* Constituents." In *Marijuana and the Cannabinoids*, edited by M. A. ElSohly, 17–49. Forensic Science And Medicine. Totowa: Humana Press. https://doi.org/10.1007/978-1-59259-947-9_2.

Brown, Horace T., and G. Harris Morris. 1893. "On Certain Functions of Hops Used in the Dry-Hopping of Beers." *Brewers Guardian*, March 28, 1893, 93–94.

Butrica, James L. 2006. "The Medicinal Use of Cannabis among the Greeks and Romans." In *Handbook of Cannabis Therapeutics: From Bench to Bedside*, edited by Ethan B. Russo and Franjo Grotenhermen, 23–42. New York: Haworth Press.

Callaway, J. C. 2004. "Hempseed as a Nutritional Resource: An Overview." *Euphytica* 140:65–72. https://doi.org/10.1007/s10681-004-4811-6.

Cherniak, Laurence. 1983. *The Great Books of Cannabis, Book II*. Oakland, CA: Damele Publishing.

Cherrett, Nia, John Barrett, Alexandra Clemett, Matthew Chadwick, and M. J. Chadwick. 2005. *Ecological Footprint and Water Analysis of Cotton, Hemp and Polyester*. Report prepared to the World Wide Fund. Stockholm Environmental Institute.

Christensen, Stephen. 2019. *HempTrain for CBD: Drastic Capital and Operating Cost Reduction by Processing Fresh/Green Hemp*. Internal report. Calgary: Canadian Greenfield Technologies.

Ciaraldi, M. 2000. "Drug Preparation in Evidence? An Unusual Plant and Bone Assemblage from the Pompeian Countryside, Italy." *Vegetation History and Archaeobotany* 9:91–98. https://doi.org/10.1007/BF01300059.

Citterio, S., A. Santagostino, P. Fumagalli, N. Prato, P. Ranalli, and S. Sgorbati. 2003. "Heavy metal tolerance and accumulation of Cd, Cr and Ni by *Cannabis sativa* L." *Plant and Soil* 256:243–252. https://doi.org/10.1023/A:1026113905129.

Clarke, R. C. 2006. "Searching for Hempen Treasures: Field Identification of Hemp Fiber in Markets, Museums and Private Collections." *Journal of Industrial Hemp* 11(2): 73–90. https://doi.org/10.1300/J237v11n02_06.

Clarke, Robert. 1977. *The Botany and Ecology of Cannabis*. California: Pods Press.

Clarke, Robert, and Mark Merlin. 2013. *Cannabis: Evolution and Ethnobotany*. Berkeley: University of California Press.

Craig, Richard B. 1980. "Operation Intercept: The International Politics of Pressure." *Review of Politics* 42(4): 556–580. https://doi.org/10.1017/S0034670500031995.

Crombie, L., and W. M. L. Crombie. 1975. "Cannabinoid formation in *Cannabis sativa* grafted inter-racially, and with two *Humulus* species." *Phytochemistry* 14(2): 409–412. https://doi.org/10.1016/0031-9422(75)85100-4.

Daniller, Andrew. 2019. "Two-thirds of Americans support marijuana legalization." Pew Research Center, November 14, 2019. https://www.pewresearch.org/fact-tank/2019/11/14/americans-support-marijuana-legalization/.

De Carcer, G. 2019. *Alcohol vs. Legal Cannabis: Consumption in North America*. Cannabis Consumer Insight Series. Washington DC: New Frontier Data.

Degenhardt, L., W. Chiu, N. Sampson, R. Kessler, J. Anthony, M. Angermeyer, R. Bruffaerts, et al. 2008. "Toward a Global View of Alcohol, Tobacco, Cannabis, and Cocaine Use: Findings from the WHO World Mental Health Surveys." *PLoS Medicine* 5(7): e141. https://doi.org/10.1371/journal.pmed.0050141.

DeLong, G. T., C. E. Wolf, A. Poklis, and A. H. Lichtman. 2010. "Pharmacological Evaluation of the Natural Constituent of Cannabis sativa, Cannabichromene and its Modulation by Δ^9-Tetrahydrocannabinol." *Drug and Alcohol Dependence* 112(1–2): 126–133. https://doi.org/10.1016/j.drugalcdep.2010.05.019.

de Meijer, E. P. M., and K. M. Hammond. 2005. "The Inheritance of Chemical Phenotype in *Cannabis sativa* L. (II): Cannabigerol Predominant Plants." *Euphytica* 145:189–198. https://doi.org/10.1007/s10681-005-1164-8.

de Meijer, Etienne. 2009. Cannabis sativa plants rich in cannabichromene and its acid, extracts thereof and obtaining extracts therefrom. GB Patent GB2459125B, filed April 10, 2008, and issued January 2, 2013.

ElSohly, M. A., and D. Slade. 2005. "Chemical Constituents of Marijuana: The Complex Mixture of Natural Cannabinoids." *Life Sciences* 78(5): 539–548. https://doi.org/10.1016/j.lfs.2005.09.011.

ElSohly, M.A., Z. Mehmedic, S. Foster, C. Gon, S. Chandra, and J.C. Church. 2016. "Changes in Cannabis Potency Over the Last Two Decades (1995-2014): Analysis of Current Data in the United States." *Biological Psychiatry* 79(7): 613–619. https://doi.org/10.1016%2Fj.biopsych.2016.01.004.

Evans, F. J. 1991. "Cannabinoids: The Separation of Central from Peripheral Effects on a Structural Basis." *Planta Medica* 57(7 Suppl.): 60–67. https://doi.org/10.1055/s-2006-960231.

Ewing, L. E., C. M. Skinner, C. M. Quick, S. Kennon-McGill, M. R. McGill, L. A. Walker, M. A. ElSohly, B. J. Gurley, and I. Koturbash. 2019. "Hepatotoxicity of a Cannabidiol-Rich Cannabis Extract in the Mouse Model." *Molecules* 24(9): 1694. https://doi.org/10.3390/molecules24091694.

Faegri, Knut, Johs. Iversen, Peter E. Kaland, and Knut Krzywinski. 1989. *Textbook of Pollen Analysis*. 4th edition. New York: John Wiley & Sons.

Finlay, D. B., K. J. Sircombe, M. Nimick, C. Jones, and M. Glass. 2020. "Terpenoids From Cannabis Do Not Mediate an Entourage Effect by Acting at Cannabinoid Receptors." *Frontiers in Pharmacology* 11:359. https://doi.org/10.3389/fphar.2020.00359.

Fischedick, J. T., A. Hazekamp, T. Erkelens, Y.H. Choi, and R. Verpoorte. 2010. "Metabolic Fingerprinting of *Cannabis sativa* L., Cannabinoids and Terpenoids for Chemotaxonomic and Drug Standardization Purposes." *Phytochemistry* 71(17–18): 2058–2073. https://doi.org/10.1016/j.phytochem.2010.10.001.

Florida Department of Health. 2012. *Harmful Algal Blooms: Economic Impacts.* Tallahassee: Florida Department of Health.

Franke, Herbert. 1974. "Siege and Defense of Towns in Medieval China." In *Chinese Ways in Warfare*, edited by F.A. Kierman and J.K. Fairbank, 151–201. Cambridge, MA: Harvard University Press.

Gabrilova, E., T. Kamada, and F. Zoutman. 2019. "Is Legal Pot Crippling Mexican Drug Trafficking Organisations? The Effect of Medical Marijuana Laws on US Crime." *Economic Journal* 129(617): 375–407. https://doi.org/10.1111/ecoj.12521.

Garcia-Jaldon, C., D. Dupeyre, and M. R. Vignon. 1998. "Fibres from Semi-retted Hemp Bundles by Steam Explosion Treatment." *Biomass and Bioenergy* 14(3): 251–260. https://doi.org/10.1016/S0961-9534(97)10039-3.

Godley, A.D., trans. 1921–25. *Herodotus*. 4 vols. London: William Heinemann.

Gray, R.D., A.J. Drummond, and S.J. Greenhill. 2009. "Language Phylogenies Reveal Expansion Pulses and Pauses in Pacific Settlement." *Science* 323(5913): 479–483. https://doi.org/10.1126/science.1166858.

Grinspoon, Lester, and James B. Bakalar. 1997. *Marihuana, the Forbidden Medicine*. Rev. ed. New Haven: Yale University Press.

Hahn, Stephen M. 2020. *Sampling Study of the Current Cannabidiol Marketplace to Determine the Extent That Products are Mislabeled or Adulterated*. Report to the U.S. House and Senate Committees on Appropriations. Washington DC: U.S. Food and Drug Administration.

Haney, A., and B. B. Kutscheid. 1975. "An Ecological Study of Naturalized Hemp (*Cannabis sativa* L.) in East-Central Illinois." *American Midland Naturalist* 93(1): 1–24. https://doi.org/10.2307/2424101.

Hanuš, L. O., S. M. Meyer, E. Muñoz, O. Taglialatela-Scafati, and G. Appendino. 2016. "Phytocannabinoids: A Unified Critical Inventory." *Natural Product Reports* 33:1357–1392. https://doi.org/10.1039/C6NP00074F.

Hauser, D. G., K. R. Van Simaeys, S. R. Lafonaine, and T. H. Shellhammer. 2019. "A Comparison of Single-Stage and Two-Stage Dry-Hopping Regimes." *Journal of the American Society of Brewing Chemists* 77(4): 251–260. https://doi.org/10.1080/03610470.2019.1668230.

Hazekamp, A., K. Bastola, H. Rashidi, J. Bender, and R. Verpoorte. 2007. "Cannabis Tea Revisited: A Systematic Evaluation of the Cannabinoid Composition of Cannabis Tea." *Journal of Ethnopharmacology* 113(1): 85–90. https://doi.org/10.1016/j.jep.2007.05.019.

Hazekamp, Arno, Justin T. Fischedick, Mónica Llano Díez, Andrea Lubbe, and Renee L. Ruhaak. 2010. "Chemistry of Cannabis." In *Development and Modification of Bioactivity*, 1033–1084. Vol. 3 of *Comprehensive Natural Products II: Chemistry and Biology*, edited by Hung-Wen (Ben) Liu and Lew Mander. Leiden: Elsevier Science. https://doi.org/10.1016/B978-008045382-8.00091-5.

Hill, A. J., C. M. Williams, B. J. Whalley, and G. J. Stephens. 2012. "Phytocannabinoids as Novel Therapeutic Agents in CNS Disorders." *Pharmacology and Therapeutics* 133(1): 79–97. https://doi.org/10.1016/j.pharmthera.2011.09.002.

Hillig, K. W., and P. G. Mahlberg. 2004. "A Chemotaxonomic Analysis of Cannabinoid Vaiation in *Cannabis* (Cannabaceae)." *American Journal of Botany* 91(6): 966–975. https://doi.org/10.3732/ajb.91.6.966.

Hsu, C. 1978. "Agricultural Intensification and Marketing Agrarianism in the Han Dynasty." In *Ancient China: Studies in Early Civilization*, by D.T. and Tsien, T. Roy, 263–264. Hong Kong: Chinese University.

Huang, L., G. Krigsvoll, F. Johansen, Y. Liu, and X. Zhang. 2017. "Carbon Emission of Global Construction Sector." *Renewable and Sustainable Energy Reviews* 81(2): 1906–1916. https://doi.org/10.1016/j.rser.2017.06.001.

Hubbard, Chase. 2020. *2020 U.S. Hemp Crop Outlook*. Trade Report. Boulder: The Jacobsen.

Hull, G. 2008. "Olive Oil Addition to Yeast as an Alternative to Wort Aeration." *Master Brewers's Association of the Americas Technical Quarterly* 45(1): 17–23.

Ip, K., and A. Miller. 2012. "Life Cycle Greenhouse Gas Emissions of Hemp–Lime Wall Constructions in the UK." *Resources, Conservation and Recycling* 69:1–9. https://doi.org/10.1016/j.resconrec.2012.09.001.

Iwata, N., and S. Kitanaka. 2011. "New Cannabinoid-Like Chromane and Chromene Derivatives from *Rhododendron anthopogonoides*." *Chemical and Pharmaceutical Bulletin* 59(11): 1409–1412. https://doi.org/10.1248/cpb.59.1409.

Izzo, A. A., F. Borrelli, R. Capasso, V. Di Marzo, and R. Mechoulam. 2009. "Non-Psychotropic Plant Cannabinoids: New Therapeutic Opportunities from an Ancient Herb." *Trends in Pharmacological Sciences* 30(10): 515–527. https://doi.org/10.1016/j.tips.2009.07.006.

King, A. J., and J. R. Dickinson. 2003. "Biotransformation of Hop Aroma Terpenoids by Ale and Lager Yeasts." *FEMS Yeast Research* 3(1): 53–62. https://doi.org/10.1016/S1567-1356(02)00141-1.

Kirkendall, J. A., C. A. Mitchell, and L. R. Chadwick. 2018. "The Freshening Power of Centennial Hops." *Journal of the American Society of Brewing Chemists* 76(3): 178–184. https://doi.org/10.1080/03610470.2018.1469081.

Kirkpatrick, K. R., and T. H. Shellhammer. 2018. "Evidence of Dextrin Hydrolyzing Enzymes in Cascade Hops (*Humulus lupulus*)." *Journal of Agricultural Food Chemistry* 66(34): 9121–9126. https://doi.org/10.1021/acs.jafc.8b03563.

Lachenmeier, D.W., and J. Rehm. 2015. " Comparative Risk Assessment of Alcohol, Tobacco, Cannabis and Other Illicit Drugs Using the Margin of Exposure Approach." *Scientific Reports* 5:8126. https://doi.org/10.1038/srep08126.

Laws, D. R. J., T. L. Peppard, F. R. Sharpe, and J. A. Pickett. 1978. "Recent Developments in Imparting Hop Character to Beer." *Journal of the American Society of Brewing Chemists* 36(2): 69–72. https://doi.org/10.1094/ASBCJ-36-0069.

Lee, Martin. 2012. *Smoke Signals: A Social History of Marijuana—Medical, Recreational, and Scientific*. New York: Simon and Schuster Press.

Li, H-L. 1973. "An Archaeological and Historical Account of Cannabis in China." *Economic Botany* 28:437–448. https://doi.org/10.1007/BF02862859.

Li, Hui-Lin. 1975. "The Origin and Use of Cannabis in Eastern Asia: Their Linguistic-Cultural Implications." In *Cannabis and Culture*, edited by Vera Rubin, 51–62. World Anthropology. The Hague: De Gruyter Mouton. https://doi.org/10.1515/9783110812060.51.

Li, X., S. Wang, G. Du, Z. Wu, and Y. Gong. 2014. "Manufacturing Particleboard Using Hemp Shiv and Wood Particles Using Low Free Formaldehyde Emission Urea-Formaldehyde Resin." *Journal of Forest Products* 64(5–6): 187–191. https://doi.org/10.13073/FPJ-D-13-00073.

LimeTechnology. 2006. *Case Study: Adnams Brewery, Reydon, NR Southwold, Suffolk*. Abingdon, Oxon.: LimeTechnology.

Linger, P., J. Müssig, H. Fischer, and J. Kobert. 2002. "Industrial Hemp (*Cannabis sativa* L.) Growing on Heavy Metal Contaminated Soil: Fibre Quality and Phytoremediation Potential." *Industrial Crops and Products* 16(1): 33–42. https://doi.org/10.1016/S0926-6690(02)00005-5.

Long, J. A., B. M. Rankin, and D. Ben-Amotz. 2015. "Micelle Structure and Hydrophobic Hydration." *Journal of the American Chemical Society* 137(33): 10809–10815. https://doi.org/10.1021/jacs.5b06655.

Lucas, P. 2017. "Rationale for Cannabis-Based Interventions in the Opioid Overdose Crisis." *Harm Reduction Journal* 14:58. https://doi.org/10.1186/s12954-017-0183-9.

Luo, X., M. A. Reiter, L. d'Espaux, J. Wong, C. M. Denby, A. Lechner, Y. Zhang, et al. 2019. "Complete Biosynthesis of Cannabinoids and Their Unnatural Analogues in Yeast." *Nature* 567(7746): 123–126. https://doi.org/10.1038/s41586-019-0978-9.

Lynch, R.C., D. Vergara, S. Tittes, K. White, C.J. Schwartz, M.J. Gibbs, T.C. Ruthenburg, K. deCesare, D.P. Land, and N.C. Kane. 2016. "Genomic and Chemical Diversity in *Cannabis*." *Critical Reviews in Plant Sciences* 35(5–6): 349–363. https://doi.org/10.1080/07352689.2016.1265363.

Manguin, Pierre-Yves. 2016. "Austronesian Shipping in the Indian Ocean: From Outrigger Boats to Trading Ships." In *Early Exchange between Aftrica and the Wider Indian Ocean World*, edited by Gwyn Campbell, 51–76. Palgrave MacMillan.

Mark, T., J. Shepherd, D. Olson, W. Snell, S. Proper, and S. Thornsbury. 2020. *Economic Viability of Industrial Hemp in the United States: A Review of State Pilot Programs*. Economic Research Bulletin. Washington DC: United States Department of Agriculture.

Masson-Delmotte, V., P. Zhai, A. Pirani, S.L. Connors, C. Péan, S. Berger, N. Caud, et al., eds. 2021. *Summary for Policymakers*. In: *Climate Change 2021: The Physical Science Basis; Contribution of Working Group I to the Sixth Assessment Report of the Intergovernmental Panel on Climate Change*. Cambridge University Press.

McClements, David J. 2015. *Nanoparticle- and Microparticle-Based Delivery Systems: Encapsulation, Protection and Release of Active Compounds*. Boca Raton: CRC Press.

McPartland, J.M. 1996a. "A review of *Cannabis* diseases." *Journal of the International Hemp Association* 3(1): 19–23. http://internationalhempassociation.org/jiha/iha03111.html.

McPartland, J.M. 1996b. "*Cannabis* Pests." *Journal of the Internation Hemp Association* 3(2): 52–55. http://internationalhempassociation.org/jiha/iha03201.html.

McPartland, J.M. 2018. "*Cannabis* Systematics at the Levels of Family, Genus, and Species." *Cannabis and Cannabinoid Research* 3(1): 203–212. https://doi.org/10.1089/can.2018.0039.

McPartland, J. M., and G.W. Guy. 2004. "The Evolution of Cannabis and Coevolution with the Cannabinoid Receptor—A Hypothesis." In *The Medicinal Uses of Cannabis and Cannabinoids*, edited by G.W. Guy, B.A. Whittle, and P. Robson, 71–101. London: Pharmaceutical Press.

McPartland, J. M., W. Hegman, and T. Long. 2019. "Cannabis in Asia: Its Center of Origin and Early Cultivation, Based on a Synthesis of Subfossil Pollen and Archaeobotanical Studies." *Vegetation History and Archaeobotany* 28:691–702. https://doi.org/10.1007/s00334-019-00731-8.

McPartland, J. M., and Z. Sheikh. 2018. "A Review of *Cannabis sativa* Based Insecticides, Miticides, and Repellents." *Journal of Entomology and Zoology Studies* 6(6): 1288–1299.

McPartland, J. M., and E. Small. 2020. "A classification of endangered high-THC cannabis (*Cannabis sativa* subsp. *indica*) domesticates and their wild relatives." *PhytoKeys* 144:81–112. https://doi.org/10.3897/phytokeys.144.46700.

Mechoulam, R., and S. Ben-Shabat. 1999. "From *gan-zi-gun-nu* to Anandamide and 2-Arachidonoylglycerol: The Ongoing Story of Cannabis." *Natural Products Reports* 16(2): 131–143. https://doi.org/10.1039/a703973e.

Mechoulam, R. 2005. "Plant cannabinoids: a neglected pharmacological treasure trove." *British Journal of Pharmacology* 146(7): 913–915. https://doi.org/10.1038/sj.bjp.0706415.

Mediavilla, V., and S. Steinemann. 1997. "Essential oil of *Cannabis sativa* L. strains." *Journal of the International Hemp Association* 4(2): 80–82. http://internationalhempassociation.org/jiha/jiha4208.html.

Mediavilla, V., M. Jonquera, I. Schmid-Slembrouck, and A. Soldati. 1998. "Decimal code for growth stages of hemp (*Cannabis sativa* L.)." *Journal of the International Hemp Association* 5(2): 65, 68–74. http://www.internationalhempassociation.org/jiha/jiha5201.html.

Meier, C. and Mediavilla, V. 1998. "Factors Influencing the Yield and Quality of Hemp (*Cannabis sativa* L.) Essential Oil." *Journal of the International Hemp Association* 5(1): 16–20. http://www.internationalhempassociation.org/jiha/jiha5107.html.

Mercuri, A.M., C.A. Accorsi, and M. Bandini Mazzanti. 2002. "The Long History of *Cannabis* and Its Cultivation by the Romans in Central Italy, Shown by Pollen Records from Lago Albano and Lago di Nemi." *Vegetation History and Archaeobotany* 11:263–276. https://doi.org/10.1007/s003340200039.

Mohan Ram, H. Y., and R. Sett. 1982. "Induction of Fertile Male Flowers in Genetically Female *Cannabis sativa* Plants by Silver Nitrate and Silver Thiosulphate Anionic Complex." *Theoretical and Applied Genetics* 62:369–375. https://doi.org/10.1007/BF00275107.

Montford, S., and E. Small. 1999. "A Comparison of the Biodiversity Friendliness of Crops with Special Reference to Hemp (*Cannabis sativa* L.)." *Journal of the International Hemp Association* 6(2): 53–63. https://www.druglibrary.org/olsen/hemp/iha/jiha6206.html.

Moonjai, N., K. Verstrepen, F. Delvaux, G. Derdelinckx, and H. Verachtert. 2002. "The Effects of Linoleic Acid Supplementation of Cropped Yeast on its Subsequent Fermentation Performance and Acetate Ester Synthesis." *Journal of the Institute of Brewing* 108(2): 227–235. https://doi.org/10.1002/j.2050-0416.2002.tb00545.x.

Naraine, S., and E. Small. 2016. "Expansion of Female Sex Organs in Response to Prolonged Virginity in *Cannabis sativa* (Marijuana)." *Genetic Resources and Crop Evolution* 63:339–348. https://doi.org/10.1007/s10722-015-0253-3.

Nigam, R. K., M. Varkey, and D. E. Reuben. 1981. "Irradiation Induced Changes in Flower Formation in *Cannabis sativa* L." *Biologia Plantarum* 23:389–391. https://doi.org/10.1007/BF02877422.

Nikvash, N., R. Kraft, A. Kharazipour, and M. Euring. 2010. "Comparative Properties of Bagasse, Canola and Hemp Particle Boards." *European Journal of Wood and Wood Products* 68:323–327. https://doi.org/10.1007/s00107-010-0465-3.

O'Brien, C., and H.S. Arathi. 2019. "Bee Diversity and Abundance on Flowers of Industrial Hemp (*Cannabis sativa* L.) ." *Biomass and Bioenergy* 122:331–335. https://doi.org/10.1016/j.biombioe.2019.01.015.

Ohlsson, A., J. E. Lindgren, A. Whalen, S. Agurell, L.E. Hollister, and H. K. Gillespie. 1980. "Plasma delta-9-Tetrahydrocannabinol Concentrations and Clinical Effects after Oral and Intravenous Administration and Smoking." *Clinical Pharmacology and Therapeutics* 28(3): 409–416. https://doi.org/10.1038/clpt.1980.181.

Okazaki, H., M. Kobayashi, A. Momohara, S. Eguchi, T. Okamoto, S. Yanagisawa, S. Okubo, and J. Kiyonaga. 2011. "Early Holocene Coastal Environment Change Inferred from Deposits at Okinoshima Archeological Site, Boso Peninsula, Central Japan." *Quaternary International* 230(1–2): 87–94. https://doi.org/10.1016/j.quaint.2009.11.002.

Page, Jonathan E., and Jana Nagel. 2006. "Biosynthesis of Terpenophenolic Metabolites in Hop and Cannabis." In *Integrative Plant Biochemistry*, edited by J. T. Romeo, 179–210. Recent Advances in Phytochemistry, vol. 40. Kidlington, Oxon.: Elsevier. https://doi.org/10.1016/S0079-9920(06)80042-0.

Pakarinen, A., J. Zhang, T. Brock, P. Maijala, and L. Viikari. 2012. "Enzymatic Accessibility of Fiber Hemp is Enhanced by Enzymatic or Chemical Removal of Pectin." *Bioresource Technology* 107:275–281. https://doi.org/10.1016/j.biortech.2011.12.101.

Pan, J. 1983. "The Invention and Development of Papermaking." In *Ancient China's Technology and Science*, compiled by the Institute of the History of Natural Sciences and Chinese Academy of Sciences, 176–183. Beijing: Foreign Languages Press.

Pate, David W. 1999. "Hemp Seed: A Valuable Food Source." In *Advances in Hemp Research*, edited by Paoli Ranalli, 243–255. London: Haworth Press.

Perez-Reyes, M. 1990. "Marijuana Smoking: Factors That Influence the Bioavailability of Tetrahydrocannabinol." *NIDA Research Monograph* 99:42–62.

Petri, G. 1988. "*Cannabis sativa*: In Vitro Production of Cannabinoids." In *Medicinal and Aromatic Plants I*, edited by Y. P. S. Bajaj, 333–349. Biotechnology in Agriculture and Forestry, vol. 4. Berlin: Springer. https://doi.org/10.1007/978-3-642-73026-9_18.

Potter, David. 2009. "The Propagation, Characterisation and Optimisation of *Cannabis sativa* L. as a Phytopharmaceutical." PhD diss., King's College London, March 2009.

Radwan, M. M., M. A. Elsohly, D. Slade, S. A. Ahmed, L. Wilson, A. T. El-Alfy, I. A. Khan, and S. A. Ross. 2008. "Non-cannabinoid Constituents from a High Potency *Cannabis sativa* Variety." *Phyochemistry* 69(14): 2627–2633. https://doi.org/10.1016/j.phytochem.2008.07.010.

Rainey, Rebecca. 2019. "New problem for legal weed: Exploding pot factories." *Politico*, February 18, 2019, 8:01 a.m. EST. https://www.politico.com/story/2019/02/18/marijuana-factories-explosions-safety-issues-1155850.

Razdan, R. K. 1986. "Structure-Activity Relationships in Cannabinoids." *Pharmacological Reviews* 38(2): 75–146.

Reglitz, K., N. Lemke, S. Hanke, and M. Steinhaus. 2018. "On the Behavior of the Important Hop Odorant 4-Mecapto-4-methylpentan-2-one (4MMP) during Dry Hopping and during Storage of Dry Hopped Beer." *BrewingScience* 71(November/December): 96–99.

Rettberg, N., M. Biendl, and L-A. Garbe. 2018. "Hop Aroma and Hoppy Beer Flavor: Chemical Backgrounds and Analytical Tools—A Review." *Journal of the American Society of Brewing Chemists* 76(1): 1–20. https://doi.org/10.1080/03610470.2017.1402574.

Robinson, Rowan. 1996. *The Great Book of Hemp*. Rochester: Park Street Press.

Rodriguez de Fonseca, R., I. Del Arco, F. J. Bermudez-Silva, A. Bilbao, A. Cippitelli, and M. Navarro. 2005. "The Endocannabinoid System: Physiology and Pharmacology." *Alcohol and Alcoholism* 40(1): 2–14. https://doi.org/10.1093/alcalc/agh110.

Roland, A., S. Delpech, and L. Dagan. 2017. "A Powerful Analytical Indicator to Drive Varietal Thiols Release in Beers: The 'Thiol Potency'." *BrewingScience* 70(November/December): 170–175. https://doi.org/10.23763/BrSc17-19roland.

Romano, L. L., and A. Hazekamp. 2013. "Cannabis Oil: Chemical Evaluation of an Upcoming Cannabis-Based Medicine." *Cannabinoids* 1(1): 1–11.

Rorabaugh, W. J. 1991. "Alcohol in America." *OAH Magazine of History*, Fall 1991, 17–19.

Ross, S. A., and M. A. ElSohly. 1996. "The Volatile Oil Composition of Fresh and Air-Dried Buds of *Cannabis sativa*." *Journal of Natural Products* 59(1): 49–51. https://doi.org/10.1021/np960004a.

Ross, S. A., Z. Mehmedic, T. P. Murphy, and M. A. ElSohly. 2000. "GC-MS Analysis of the Total delta-9 THC Content of Both Drug- and Fiber-Type Cannabis Seeds." *Journal of Analytical Toxicology* 24(8): 715–717. https://doi.org/10.1093/jat/24.8.715.

Rudgley, Richard. 1995. *Essential Substances: A Cultural History of Intoxicants in Society*. New York: Kodansha.

Russo, E. B. 2002. "Cannabis Treatments in Obstetrics and Gynecology: A Historical Review." *Journal of Cannabis Therapeutics* 2(3–4): 5–35. https://doi.org/10.1300/J175v02n03_02.

Russo, E. B. 2007. "History of Cannabis and Its Preparations in Saga, Science, and Sobriquet." *Chemistry and Biodiversity* 4(8): 614–648. https://doi.org/10.1002/cbdv.200790144.

Russo, E. B. 2011. "Taming THC: Potential Cannabis Synergy and Phytocannabinoid-Terpenoid Entourage Effects." *British Journal of Pharmacology* 163(7): 1344–1364. https://doi.org/10.1111/j.1476-5381.2011.01238.x.

Russo, E., H-E. Jiang, X. Li, A. Sutton, A. Carboni, F. del Bianco, G. Mandolino, et al. 2008. "Phytochemical and Genetic Analyses of Ancient Cannabis from Central Asia." *Journal of Experimental Botany* 59(15): 4171–4182. https://doi.org/10.1093/jxb/ern260.

Russo, E. B., and G. W. Guy. 2006. "A Tale of Two Cannabinoids: The Therapeutic Rationale for Combining Tetrahydrocannabinol and Cannabidiol." *Medical Hypotheses* 66(2): 234–246. https://doi.org/10.1016/j.mehy.2005.08.026.

Russo, Ethan B., and Jahan Marcu. 2017. "Cannabis Pharmacology: The Usual Suspects and a Few Promising Leads." In *Cannabinoid Pharmacology*, edited by D. Kendall and S. P. H. Alexander, 67–134. Advances in Pharmacology, vol. 80. Academic Press.

Santiago, M., S. Sachdev, J. C. Arnold, I. S. McGregor, and M. Connor. 2019. "Absence of Entourage: Terpenoids Commonly Found in *Cannabis sativa* Do Not Modulate the Functional Activity of Δ^9-THC at Human CB_1 and CB_2 Receptors." *Cannabis and Cannabinoid Research* 4(3): 165–176. https://doi.org/10.1089/can.2019.0016.

Schlosser, Eric. 2003. *Reefer Madness: Sex, Drugs, and Cheap Labor in the American Black Market*. Boston: Houghton Mifflin.

Schöpper, C., A. Kharazipour, and C. Bohn. 2009. "Production of Innovative Hemp Based Three-Layered Particleboards with Reduced Raw Densities and Low Formaldehyde Emission." *International Journal of Materials and Product Technology* 36(1–4): 358–371. https://doi.org/10.1504/IJMPT.2009.027842.

Schultes, R. E. 1970. "Random Thoughts and Queries on the Botany of *Cannabis*." In *The Botany and Chemisty of* Cannabis, edited by C. R. B. Joice and S. H. Curry, 11–38. London: J. & A. Churchill.

Schultes, R.E., A. Hofmann, and C. Rätsch. 2001. *Plants of the Gods: Their Sacred, Healing, and Hallucinogenic Powers*. 2nd ed. Rochester: Healing Arts.

Severino, P., T. Andreani, A. S. Macedo, J. F. Fangueiro, M. H. A. Santana, A. M. Silva, and E. B. Souto. 2012. "Current State-of-Art and New Trends on Lipid Nanoparticles (SLN and NLC) for Oral Drug Delivery." *Journal of Drug Delivery* 2012:750891. https://doi.org/10.1155/2012/750891.

Shapira, A., P. Berman, K. Futoran, O. Guberman, and D. Meiri. 2019. "Tandem Mass Spectrometric Quantification of 93 Terpenoids in Cannabis Using Static Headspace Injections." *Analytical Chemistry* 91(17): 11425–11432. https://doi.org/10.1021/acs.analchem.9b02844.

Sharma, G. K. 1977. "Cannabis Folklore in the Himalayas." *Botanical Museum Leaflets* (Harvard University) 25(7): 203–215.

Sherratt, Andrew. 1995. "Alcohol and Its Alternatives: Symbol and Substance in Pre-Industrial cultures." In *Consuming Habits: Drugs in History and Anthropology*, edited by J. Goodman, P. Lovejoy, and A. Sherratt, 11–46. London: Routledge.

Sherratt, Andrew. 1997. *Economy and Society in Prehistoric Europe: Changing Perspectives*. Princeton University Press.

Small, E. 2015. "Response to the Erroneous Critique of my *Cannabis* Monograph by R. C. Clarke and M.D. Merlin." *Botanical Review* 81:306–316. https://doi.org/10.1007/s12229-015-9159-1.

Small, Ernest. 2017. *Cannabis: A Complete Guide*. Boca Raton: CRC Press.

Small, E., and T. Antle. 2003. "A Preliminary Study of Pollen Dispersal in *Cannabis sativa* in Relation to Wind Direction." *Journal of Industrial Hemp* 8(2): 37–50. https://doi.org/10.1300/J237v08n02_03.

Small, E., and S. Naraine. 2016. "Expansion of female sex organs in response to prolonged virginity in *Cannabis sativa* (marijuana)." *Genetic Resources and Crop Evolution* 63:339–348. https://doi.org/10.1007/s10722-015-0253-3.

Steyer, D., P. Tristram, C. Clayeux, F. Heitz, and B. Laugel. 2017. "Yeast Strains and Hop Varieties Synergy on Beer Volatile Compounds." *BrewingScience* 70(September/October): 131–141. https://doi.org/10.23763/BrSc17-13Steyer.

Stokes, J. R., R. Hartel, L. B. Ford, and T. B. Casale. 2000. "Cannabis (Hemp) Positive Skin Tests and Respiratory Symptoms." *Annals of Allergy, Asthma and Immunology* 85(3): 238–240. https://doi.org/10.1016/s1081-1206(10)62473-8.

Stonehous, G. C., B. J. McCarron, Z. S. Guignardi, A. F. El Mehdawi, L. W. Lima, S. C. Fakra, and E. A. H. Pilon-Smits. 2020. "Selenium Metabolism in Hemp (*Cannabis sativa* L.)—Potential for Phytoremediation and Biofortification." *Environmental Science and Technology* 54(7): 4221–4230. https://doi.org/10.1021/acs.est.9b07747.

Takoi, K., Y. Itoga, K. Koie, T. Kosugi, M. Shimase, Y. Katayama, Y. Nakayama, and J. Watari. 2010. "The Contribution of Geraniol Metabolism to the Citrus Flavour of Beer: Synergy of Geraniol and β-Citronellol Under Coexistence with Excess Linalool." *Journal of the Institute of Brewing* 116(3): 251–260. https://doi.org/10.1002/j.2050-0416.2010.tb00428.x.

Takoi, K., Y. Itoga, J. Takayanagi, I. Matsumoto, and Y. Nakayama. 2016. "Control of Hop Aroma Impression of Beeer with Blend-Hopping using Geraniol-Rich Hop and New Hypothesis of Synergy among Hop-derived Flavour Compounds." *BrewingScience* 69(November/December): 85–93.

Tomke, Prerana D., and Virenda K. Rathod. 2020. "Nanoengineering Tools in Beverage Industry." In *Nanoengineering in the Beverage Industry*, edited by A. M. Grumezescu and A. M. Holban, 35–69. Science of Beverages, vol. 20. Duxford: Woodhead Publishing. https://doi.org/10.1016/B978-0-12-816677-2.00002-8.

USDA. 2021. "Irrigation & Water Use." Updated August 27, 2021. https://www.ers.usda.gov/topics/farm-practices-management/irrigation-water-use/.

van Bakel, H., J. M. Stout, A. G. Cote, C. M. Tallon, A. G. Sharpe, T. R. Hughes, and J. E. Page. 2011. "The Draft Genome and Transcriptome of *Cannabis sativa*." *Genome Biology* 12(10): R102. https://doi.org/10.1186/gb-2011-12-10-r102.

Venkatarama Reddy, B. V., and K. S. Jagadish. 2003. "Embodied Energy of Common and Alternative Building Materials and Technologies." *Energy and Buildings* 35(2): 129–137. https://doi.org/10.1016/S0378-7788(01)00141-4.

Voklow, N. D., R. D. Bahler, W. M. Compton, and S. R. B. Weiss. 2014. "Adverse Health Effects of Marijuana Use." *New England Journal of Medicine* 370:2219–2227. https://doi.org/10.1056%2FNEJMra1402309.

Wagner, H., and G. Ulrich-Merzenich. 2009. "Synergy Research: Approaching a New Generation of Phytopharmaceuticals." *Phytomedicine* 16(2–3): 97–110. https://doi.org/10.1016/j.phymed.2008.12.018.

Walker, R., and S. Pavía. 2014. "Moisture Transfer and Thermal Properties of Hemp–Lime Concretes." *Construction and Building Materials* 64:270–276. https://doi.org/10.1016/j.conbuildmat.2014.04.081.

Wang, M., Y-H. Wang, B. Avula, M. M. Radwan, A. S. Wanas, J. van Antwerp, J. F. Parcher, M. A. ElSohly, and I. A. Khan. 2016. "Decarboxylation Study of Acidic Cannabinoids: A Novel Approach Using Ultra-High-Performance Supercritical Fluid Chromatography/Photodiode Array-Mass Spectrometry." *Cannabis and Cannabinoid Research* 1(1): 262–272. https://doi.org/10.1089/can.2016.0020.

Wardle, M. C., B. A. Marcus, and H. de Wit. 2015. "A Preliminary Investigation of Individual Differences in Subjective Responses to D-Amphetamine, Alcohol, and Delta-9-Tetrahydrocannabinol Using a Within-Subjects Randomized Trial." *PLoS ONE* 10(10): e0140501. https://doi.org/10.1371/journal.pone.0140501.

Warf, B. 2014. "High Points: An Historical Geography of Cannabis." *Geographical Review* 104(4): 414–438. https://doi.org/10.1111/j.1931-0846.2014.12038.x.

Whittle, Brian Anthony. 2003. Cannabinoid liquid formulations for mucosal administration. United Kingdom Patent EP1542657A1. August 14.

WHO. 2018a. *Cannabidiol (CBD): Critical Review Report*. Report prepared for 40th Expert Committee on Drug Dependence, June 4–7, 2018. Geneva: World Health Organization.

WHO. 2018b. *Expert Committee on Drug Pependence Pre-Review: Delta-9-tetrahydrocannabinol*. Geneva: World Health Organization.

WHO. 2018c. *WHO Expert Committee on Drug Dependence: Critical Review; Cannabis and Cannabis Resin*. Report prepared for 41st Expert Committee on Drug Dependence. Geneva: World Health Organization.

Williamson, E. M. 2001. "Synergy and Other Interactions in Phytomedicines." *Phytomedicine* 8(5): 401–409. https://doi.org/10.1078/0944-7113-00060.

Windsor, H.H. 1938. "New Billion-Dollar Crop." *Popular Mechanics*, vol. 69, no. 2, February 1938.

Wood, T. B., W. T. Newton Spivey, and T. H. Easterfield. 1899. "Cannabinol. Part I." *Journal of the Chemical Society, Transactions* 75:20–36. https://doi.org/10.1039/CT8997500020.

Zablocki, B., A. Aidala, S. Hansell, and H. Raskin White. 1991. "Marijuana Use, Introspectiveness and Mental Health." *Journal of Health and Social Behavior* 32(1): 65–79. https://doi.org/10.2307/2136800.

Zhang, L.L., R.Y. Zhu, J.Y. Chen, J.M. Chen, and X.X. Feng. 2008. "Seawater-Retting Treatment of Hemp and Characterization of Bacterial Strains Involved in the Retting Process." *Process Biochemistry* 43(11): 1195–1201. https://doi.org/10.1016/j.procbio.2008.06.019.

INDEX